ARBORICULTURE AND THE LAW IN CANADA

By
Dr. Julian A. Dunster
R.P.F., M.C.I.P., ISA Certified Arborist
ASCA (candidate)
and
Susan M. Murray MSc., P.Ag
ISA Certified Arborist

BOOKS

Published by
International Society of Arboriculture
P.O. Box GG
Savoy, Illinois 61874 USA

Disclaimer

This book contains a discussion of the main issues concerning arboriculture and the law in Canada. It is intended to be a useful starting point, but it is neither an exhaustive coverage of every known case, nor is it to be used or construed as the sole source of legal advice. Since new precedents arise continuously, it can be anticipated that some of the points discussed may change with time. Neither the authors nor the International Society of Arboriculture accept any responsibility for the manner in which readers may use or interpret the contents of this book.

Julian A. Dunster
Susan Murray
Bowen Island, British Columbia. 1997

ISBN: 1-881956-19-9

Arboriculture and the Law in Canada
Julian A. Dunster and Susan M. Murray

Production and Editing: TerraMare
Graphics and Cover Design: Emily McCullum
Printed in Canada by: Benwell Atkins Ltd.
901 Great Northern Way
Vancouver, B.C. V5T 1E1

International Society of Arboriculture
Web site: http://www.ag.uiuc.edu/~isa
e-mail: isa@isa-arbor.com
10 9 8 7 6 5 4 3 2 1

Preface

Arboriculture is the science and art of growing, tending, and managing trees and shrubs for purposes other than industrial forestry. In Canada, arboriculture is rapidly evolving, both as a discipline and in its relationship to urban forestry. In the past six years, certification of arborists with the International Society of Arboriculture has become a basic requirement in many parts of the country. Training courses, trade journals, conferences, and new ideas have become more widely available. Public awareness of good tree care is now rapidly increasing, though it still has a long way to go. Other professions are routinely adopting arboricultural practices. Overall, rising expectations are imposing increasingly more stringent demands on the practising professional arborist.

As arboriculture becomes more widely recognized, high industry-wide professional standards become necessary to maintain the credibility of the discipline. Among other things, professional arborists need a basic understanding of the legal rights and obligations pertaining to the ownership of or responsibility for trees in Canada. Appearances in court will then be made for all the right reasons.

As the materials in this book demonstrate, there is no shortage of Canadian legal precedents for disputes involving trees and shrubs, especially those on private property. For instance, neighbours have cut each other's trees down with alarming regularity; trees have fallen on roads, with crippling and sometimes fatal results; traffic signs have been obscured; roots have caused damage beyond property lines; power lines have been knocked down; and trees close to power lines have been climbed.

This book contains a comprehensive collection of case law from across Canada. The case outlines and analyses in the first nine chapters give a good sense of the ways in which legal disputes involving trees have arisen and been settled. In the final chapter, aspects of professional arboricultural practice are discussed.

The materials in this book have been assembled as a basic reference for arborists, lawyers, administrators, the general public, and anyone else with an interest in trees.

The authors are always interested in hearing about new cases or interpretations to be used in subsequent editions. Suggestions for changes and updated case histories can be sent to:

Julian Dunster
Box 109
Bowen Island, British Columbia
Canada, V0N 1G0
or
c/o International Society of Arboriculture

It is intended to periodically update the book in light of new precedents.

Acknowledgements

We are indebted to the many people who have helped with the research, writing, editing, and final production of this book.

Researching and writing a book about arboriculture and the law is a challenging task; one made that much more interesting because both of the authors are practising arborists rather than lawyers. We have, therefore, been fortunate to garner the support of several lawyers who have helped to mould the legal details in this book and minimise mistakes. Christopher Lemon and articling students Volker Helmuth and Adrian Huzel at the law firm of Ferguson Gifford in Vancouver provided initial help and access to Ferguson Gifford's computer data base. Christopher Lemon also provided a critical review of the first draft and advice on legal terminology and concepts. J.M. (Tim) Mackenzie of the law firm Singleton Urquhart Scott, also in Vancouver, ensured that all the legal citations were properly written, assisted with materials and concepts for Chapter 10, and also provided insightful comments on the overall text. With the help of Christopher and Tim, this book has been greatly improved.

Many people across Canada have contributed cases and comments for inclusion in the book. These people include Gerry Chaster, Vancouver Island, B.C.; Christopher G. Nation, Debbie Marzocco, and Ron Carter, District of Saanich, B.C.; Jim Hamilton, City of Richmond, B.C.; Bruce McGauley and

Geoffrey Belch, City of London, Ontario; David Schmitt, City of Kitchener, Ontario; Ian Bruce, Thompson, Bruce & Associates, Etobicoke, Ontario; and Mary Donovan, City of Halifax, Nova Scotia.

Emily McCullum provided extremely useful editing advice and did the layout and final production. Steve Day of Day & Associates in Colorado, Christopher Lemon, Tim Mackenzie and Ian Bruce reviewed the first draft and provided direction for the final text. Kwantlen University College provided Susan Murray with a term educational leave.

The book was sponsored by a grant from the International Society of Arboriculture. We are indebted to the staff at the ISA for their assistance throughout this project.

Contents

Chapter 3: Trees Damaged During Trespass

Chapter 4: Trees Growing On Boundary Lines

Chapter 5: Trees Along Roads And Highways

Chapter 6: Trees Affecting Utility Rights-of-way and Easements

Introduction

This book is about the legal responsibilities attached to owning or working near trees. Issues addressed include how these responsibilities vary with different circumstances, including those of a neighbouring land owner (or anyone else). The emphasis in this book is on arboriculture and associated activities, rather than on commercial forest or timber management.

Arboriculture is defined in Dunster and Dunster (1996) as:

> The science and art of growing, tending, and managing trees and shrubs as individuals or in groups, typically for their value as ornaments and aesthetics, their ability to provide visual and noise screening, and climatic moderation, typically in an urban setting. Practitioners are termed arborists.

However, this book is written for use not only by arborists, but also by lawyers, staff in all levels of government (especially municipal), and the public. The authors' experience in preparing evidence for court cases suggested that this book's content should fall somewhere between a checklist of simple do's and don'ts and a lengthy discussion of legal principles. Consequently, the book includes a simplified overview of legal principles, concepts, and definitions, and a comprehensive review of the court cases (including several important early precedents from Great Britain) which define the limits of acceptable practice.

To fully comprehend the complexity and practical implications of legal interpretations, it is useful to have a basic understanding of how courts function, the legal processes that must be followed, and the meaning of certain key terms. These fundamentals are covered in Chapters 1 and 2, providing a framework for the case histories presented later on. The duties involved in owning or being responsible for trees and related basic legal principles are also covered in these initial chapters.

In each of Chapters 3 through 7, court cases dealing with specific issues, such as trespass or trees near highways, are examined. Chapter 8 contains other tree related cases involving such issues as covenants, personal injury, subdivision approvals, and trees falling across property lines. In Chapter 9, tree protection by-laws are discussed. Advice for the arborist acting as a consulting expert or an expert witness is offered in Chapter 10. These latter two chapters are specifically aimed at arborists, although parts of Chapter 10 may prove useful to other consultants.

Cases in this book were chosen to provide a general indication of judicial opinion on tree related disputes

Only some of the many tree related Canadian court cases are cited in this book. These cases were chosen to provide a general indication of judicial opinion on tree related disputes, though in several instances cases having similar facts have concluded with quite different judgements. There is no detailed analysis of how legal precedents have evolved over time, as new cases arose. Only two of the selected cases are from the Province of Quebec, partly because most are written in

French, and partly because the authors were unable to garner assistance from the arboricultural community in Quebec. It is assumed that, although the legal system in Quebec is based on a system of civil codes, the principles involved in arriving at an equitable judgement are similar to those found under English common law.

Many of the cases cited contain several important principles beyond the main issue, and are thus referenced more than once. For example, the case of *Horseshoe Bay Retirement Society v. S. I. F. Development Corp.* (1990) involved trespass, but the judgement provides guidance on how damage and damages are to be assessed, and how punitive and general damages are to be awarded. This judgement is now widely cited. In other instances, one or more parts of a judgement are more limited in application, perhaps even restricted to the province of origin. Such limitations are often seen where there is legislation, such as a Municipal Act or a Highways Act, which varies in scope and application across the country.

...the services of a lawyer are essential

Wherever possible, standard legal terms have been defined, using interpretations from precedents or from Black's Law Dictionary (6th Edition). All the cases mentioned in the text are fully referenced in the Table of Cases in the Appendix. These cases are cited using accepted legal notation to provide the reader with an accurate means of gaining further information.

While every effort has been made to provide

a comprehensive review of the relevant tree related court cases, some specific aspects, cases, or legal interpretations may be missing from the text or may change as new precedents arise. As with all such matters, when the time comes to initiate a court action, the services of a lawyer are essential.

Chapter 1
Trees and the Legal Fundamentals

1.1 Introduction

It is generally considered a property owner's responsibility to ensure that the tree or trees on the property do not pose a risk to other people or property. Responsibility implies that one or more persons have been designated to look after the tree, either because they own the land upon which the tree is growing or because they are employed directly (an employee) or indirectly (a contractual relationship) by the owner. Similarly, a person or organisation whose activities may affect the condition of a tree is assumed to be responsible for the effects of those activities. Many of the cases in this book hinge on the legal responsibilities attached to ownership and to activities that affect the condition of one or more trees.

Most of the cases cited in this book fall under the law of torts, as opposed to criminal law. The body of case law comprising the law of torts aims to protect the interests of individuals. A tort is defined by Black (1990) as:

> A private or civil wrong or injury, including action for bad faith breach of contract, for which the Court will provide a remedy in the form of an action for damages.
>
> A violation of a duty imposed by general law or otherwise upon all persons occupying the relation to each other which is involved in a given transaction.

> There must always be a violation of some duty owing to the plaintiff, and generally such duty must arise by operation of law and not by mere agreement of the parties.
>
> A legal wrong committed upon the person or property independent of contract. It may be either (1) a direct invasion of some legal right of the individual; (2) the infraction of some public duty by which special damage accrues to the individual; (3) the violation of some private obligation by which like damage accrues to the individual.

In a tree related court case, therefore, there has to be proof that a clearly identified cause (the proximate cause) led to a clearly identifiable result (typically seen as some form of damage). The reasons for the cause may vary considerably, and may include ignorance, negligence, wilful trespass, Acts of God, or an unanticipated cumulative effect causing subsequent damage.

...there has to be proof that a clearly identified cause led to a clearly identifiable result

Proximate cause is defined in Black (1990) as:

> That which, in a natural and continuous sequence, unbroken by any efficient intervening cause, produces injury, and without which the result would not have occurred.
>
> That which stands next in causation to the effect, not necessarily in time or space but in causal relation.
>
> An injury or damage is proximately caused by an act, or a failure to act, whenever it appears from the evidence in the case, that the act or omission played a substantial part in bringing about or actually causing the injury or damage; and that the injury or damage was either a direct result or a reasonably probable consequence of the act or omission.

The types of damage also vary considerably, but typically include death of one or more trees, removal of one or more trees, damage eventually requiring removal of one or more trees, or damage caused by all or part of a tree falling onto property or people. Loss of privacy, loss of aesthetic values, slope destabilisation, and loss of valuable lumber are all potential effects of such damage.

To launch a lawsuit, the person initiating it, termed the plaintiff, files a writ of summons and a statement of claim, including details of the damage caused, and a general description of the losses suffered from such damage. The person who is alleged to have caused the problem, the defendant, then has an opportunity to respond by filing a statement of defence. In many instances, the dispute is settled out of court. The cases cited throughout this book demonstrate what can happen if a court case proceeds.

1.2 The Basic Elements of a Court Case

1.2.1 Duty of Care and Standard of Care

Under English common law there is a presumption that the owner of or the person responsible for a tree has what is termed a duty of care for that tree. While the exact legal definition varies, a duty is defined in Black (1990) as:

> A human action which is exactly conformable to the laws which require us to obey them. Legal or moral obligation. An obligation that one has by law or contract. Obligation to conform to legal standard of reasonable conduct in light of apparent risk. An obligation, recognized by the

> law, requiring actor to conform to certain standard of conduct for protection of others against unreasonable risks.

Once a duty of care is established, the standard of care determines the extent to which any one individual's actions fulfil the duty. Black (1990) defines the standard of care as:

> ...that degree of care which a reasonably prudent person should exercise in same or similar circumstances. If a person's conduct falls below such standard, he may be liable in damages for injuries or damages resulting from his conduct.

Black (1990) defines a standard as:

> Stability, general recognition, and conformity to established practice. A type, model, or combination of elements accepted as correct or perfect. A measure or rule applicable in legal cases such as "standard of care in tort actions."

...the standard of care determines how the duty of care is measured

Thus, the standard of care determines how the duty of care is measured.

In general then, the owner of one or more trees has some degree of legal responsibility (the duty) to exercise common prudence (the standard) in maintaining his or her trees in such a way that they will not fall down or otherwise fail in a manner likely to cause damage to other property or people. The duty and standard of care have been extended by the courts to include, in certain circumstances, a duty to inspect the trees (or cause them to be inspected by a qualified person) according to generally accepted standards to discover whether they have any defects that may cause them to fail.

But, to establish liability, it is necessary to show that the tree owner "ought to have known" the trees' condition or "would have known if he had taken the precaution that a reasonable landowner would take" (*Hayes v. Davis* [1991]).

The standard of care is greatly increased if an unsafe condition or the likelihood of an unsafe condition has been drawn to the owner's attention and no remedial action has been taken. In *Hayes v. Davis* (1991), the case of *Leakey v. Nat. Trust for Places of Historic Interest or Natural Beauty* (1980) was examined and quoted as follows:

> A person on whose land a hazard naturally occurred, whether in the soil itself or in something growing on the land, and which encroached or threatened to encroach onto another's land thereby causing or threatening to cause damage, was under a duty, if he knew or ought to have known of the risk of encroachment, to do what was reasonable in all the circumstances to prevent or minimise the risk of the known or foreseeable damage or injury to the other person or his property.

The trial judge in *Hayes* agreed with the plaintiffs, who had expressed concerns to their neighbour about the safety of several of his trees. In the judge's opinion, this simple discussion was sufficient to establish a duty of care that was breached by the defendant's inaction—the trees subsequently fell down and injured one of the plaintiffs. An appeal was launched but dismissed with reference to *Leakey,* albeit with a dissenting judgement that, in the absence of any visible defects in the trees, merely expressing concern was insufficient grounds for establishing liability. *Hayes* is especially important because it highlights the manner in which concern about another's trees might be expressed and the manner in which the tree owner might be expected to respond; that is, the extent to which the standard is expected to be upheld.

In all such cases, the test of responsibility hinges on what a prudent or reasonable person would normally be expected to do, and what outcomes this same person might reasonably have been expected to foresee. The courts define what is reasonable and ought to be expected in ordinary circumstances. While one might expect a more relaxed standard of care for the private citizen lacking expertise in the recognition of tree hazards, a utility company or a transportation agency that could recognise the potential for hazards and bring in experts to assess them would be expected to adhere to a higher standard. The case of *Kelley v. Ingersoll (Town)* (1996) provides an example of how liability can be attached to a private landowner when both the duty and standard of care are breached (see Section 8.4 for further details on this case).

...the courts define what is reasonable

Many factors affect the duty and standard of care themes. For example, did the tree grow naturally or was it planted (natural versus human intervention)? Did failure of the tree or part of the tree occur due to Acts of God or as a result of human error?

Black (1990) defines an Act of God as:

> An Act occasioned exclusively by forces of nature without the interference of any human

> agency. A misadventure or casualty is said to be caused by the "act of God," when it happens by the direct, immediate, and exclusive operation of the forces of nature, uncontrolled and uninfluenced by the power of man, and without human intervention, and is of such a character that it could not have been prevented or escaped from by any amount of foresight or prudence, or by any reasonable degree of care or diligence, or by the aid of any appliances which the situation of the party might reasonably require him to use. Any accident produced by any physical cause which is irresistible, such as lightning, tempests, perils of the sea, tornadoes, earthquakes.

Could the incident in question have been reasonably foreseen? Did any government own or bear responsibility for the trees in question and did it inspect them? If so, was the inspection carried out correctly? The latter question is of special interest and arises in several tree related cases.

Court precedents are not solely restricted to ownership. The duty of care also extends to a person or organisation whose activities might cause damage to one or more trees or result in subsequent damage to a third party. Thus, for example, electricity lines passing through trees that might be climbed by children or electricity lines near defective trees have been designated as a danger by the courts, and electrical utility companies are now charged with an extraordinary duty of care in managing power lines and trees (discussed in detail in Section 6.3). Similarly, anyone undertaking work that might affect a tree located on another property has a duty of care to consider the outcome of their actions.

If a clearly defined cause can be proven to have led to a clearly assessable effect (the resulting damage), then the issue of negligence may arise.

1.2.2 Negligence

Negligence takes many forms. Black (1990) defined it as:

> The omission to do something which a reasonable man, guided by those ordinary considerations which ordinarily regulate human affairs, would do, or the doing of something which a reasonable and prudent man would not do.
>
> Negligence is the failure to use such care as a reasonably prudent and careful person would use under similar circumstances; it is the doing of some act which a person of ordinary prudence would not have done under similar circumstances or failure to do what a person of ordinary prudence would have done under similar circumstances.
>
> It [negligence] is characterized chiefly by inadvertence, thoughtlessness, inattention, and the like, while "wantonness" or "recklessness" is characterized by wilfulness.
>
> [The] Doctrine of Negligence rests on a duty of every person to exercise due care in his conduct toward others from which injury may result.

In general terms, the actions necessary to establish negligence require proof that a duty of care existed, that this duty of care has been breached in some manner, and that damage has resulted from this breach. Wright et al (1985) discussed the issue of negligence in detail.

They adopted a six part scheme to determine negligence:

1) the defendant's conduct must be negligent, that is, in breach of the standard of care set by the law;

2) the claimant must suffer some damage;

3) the damage suffered must be caused by the negligent conduct of the defendant;

4) there must be a duty recognized by the law to avoid this damage;

5) the conduct of the defendant must be a proximate cause of loss or, stated in another way, the damage should not be too remote a result of the defendant's conduct;

6) the conduct of the plaintiff should not be such as to bar his recovery, that is, he must not be guilty of contributory negligence and he must not voluntarily assume the risk.

Several forms of negligence are seen in tree related cases, including:

- **Ordinary Negligence**
  A simple omission or want of what would be considered ordinary care
- **Passive Negligence**
  An omission of what should have been done—that is, the problem or defect was known but not fixed
- **Negligence Per Se**
  A presumption of negligence without specific proof because the injurious act violated a public statute or is so wanton or reckless as to create the presumption of negligence
- **Contributory Negligence**
  Where the plaintiff's negligence contributes to the injury or damage for which he or she

has suffered and is partially responsible

- **Gross Negligence**
  An intentional failure to meet accepted standards, with no regard for the likely consequences, even though they were known at the time
- **Criminal Negligence**
  A flagrant disregard for or wilful indifference to the safety of other people
- **Res Ipsa loquitur**
  Where the occurrence of an injurious act in itself gives rise to an inference (rather than a presumption) of negligence, thereby putting the onus on the defendant to provide an alternative explanation

In all cases, if negligence can be proved, the negligent party may be exposed to some degree of liability. The degree of negligence affects the amount of liability. The two crucial factors in assessing negligence are the magnitude of the risks and the likelihood of injury. Note that once it has been proven that a party has a duty of care, perhaps through a by-law or by repeatedly taking responsibility for a certain action, that party becomes responsible for meeting the standard of care. There are several cases where a default duty of care has been established, and liability in negligence has ensued (for example, *Crosty v. Burlington[City]* [1978] and *Huestis v. Toronto [City]* [1926]; see Sections 4.3.1 and 5.3).

Liability or legal responsibility determines who pays, and how much. In general terms, liability will be determined by whether there was a duty of care to be exercised, and if there was, whether someone was negligent in exercising that duty of care. If it can be shown that a duty of care existed and was not upheld as a result of negligence, the person(s) at fault will likely be liable for the resulting damages and associated costs. It should be noted that there is a difference between a person directly causing a problem (an act of commission) and a person who fails to prevent or remove a problem (an act of omission).

Liability can take several forms, including:

- **Contingent Liability**
  A liability not yet determined but dependent on events yet to take place
- **Joint Liability**
  Where more than one person is clearly responsible (often the landowner and contractor)
- **Primary Liability**
  Where one person is clearly identified as being directly responsible
- **Secondary Liability**
  Liability triggered when certain primary actions take place
- **Legal Liability**
  The liability amount imposed in a judgement
- **Professional Liability**
  The professional involved makes certain pronouncements and accepts legal responsibility for their consequences

- **Vicarious Liability**
  The imposition of liability on one person for the actionable conduct of another, based solely on a relationship between the two persons. Indirect or imputed legal responsibility for acts of another; for example, the liability of an employer for the acts of an employee.

Liability issues become more complex when any level of government is involved, since liability may be affected by relevant legislation or by a prior policy decision. If legislation, for instance, dictates certain actions, it may be used to absolve liability. Similarly, the failure of government or other public authority to act in light of known defects may attract liability, unless there has been a prior policy decision to do nothing due to other constraints. Examples of situations where liability may be complicated in this way include traffic safety signs obscured by trees, potentially hazardous trees on a highway right-of-way, and trees with potential to affect a right-of-way.

This split between policy and operational activities emanates from the British decision *Anns v. Merton London Borough Council* (1978) and has since been applied in important test cases in the Supreme Court of Canada in *Just v. British Columbia* (1989) and *Swinamer v. Nova Scotia (Attorney-General)* (1992). In essence, the prevailing ruling in Canadian cases is that public authorities have a duty of care to take reasonable steps to ensure the safety of people using the highways. But, and it is has proven to be a very important caveat, there may be no duty of care attached to a public authority when it can be clearly demonstrated that an incident occurred, but the duty is limited as a result of a policy decision (decisions which involve or are dictated by financial, economic, social or political factors or constraints). For example, if the responsible authority had made a policy decision to inspect highways for hazard trees and implemented the policy within the limits of budget and staff time, then any new incident in an area not yet inspected might be exempt from liability. The courts have ruled that the general problem had been recognized and was being addressed, though the process was constrained by time and money. However, if it could be shown that the incident involved an area already inspected (in accordance with the policy decision), then the case would hinge on what was inspected, how well, and with what degree of rigour. If the inspection process (the implementation stage of the policy decision) was flawed, then it might be possible to show negligence and liability might follow.

...there may be no duty of care attached to a public authority when...the duty is limited as a result of a policy decision

Liability on any one parcel of property may be covered by an Occupiers Liability Act under which the occupier, the person in charge of or having responsibility for the premises, has a duty of care to ensure that said premises are safe for others to enter. For example, *Choppin v. Caouette* (1982) involved a case where a man visited a friend's cabin, stepped off the porch, got his leg caught by a tree root and fell over, breaking his leg. On previous occasional visits there had been a makeshift step at the place where he stepped off. The Court found the cabin owner negligent in not continuing to maintain a step over the root in question, thereby causing the accident. The Court also found the plaintiff guilty of contributory negligence in not exercising reasonable care as he stepped off the porch (that is, he did not look where he was going). Once the damages for loss of mobility, pain, suffering, and loss of income were established, the Court then awarded 50% of these damages, since the plaintiff was held to be 50% responsible for the accident.

In specific instances, it may be feasible to enter into a contract under an Occupiers Liability Act whereby one or more parties may modify the occupier's duty of care. However, such a contract would apply only to the parties identified in the contract, which can lead to problems should another party be affected. For example, an owner who wishes to remove a tree may be asked not to by someone who offers to accept liability for the tree. Should an agreement to that effect be reached, problems may arise if the tree damages a third party who had no knowledge of the agreement regarding liability. This problem arises in the protection of environmentally sensitive areas which encompass trees (such as a watercourse with tree lined fringes). Can provincial or federal environment agencies (or other ministries) force a landowner to retain trees that, in the

problems may arise if a tree damages a third party who had no knowledge of an agreement regarding liability

opinion of his or her expert, are hazardous? The answer appears to be no, since in the event of damage, the owner of the trees would still be liable, and would then have to sue the other agency for damages.

Another complicated question of liability arises when an independent contractor causes incidental damage. In the case of *Milliken v. Kidd and Kidd* (1994), the contractor proved to the satisfaction of the judge that he was acting on instructions from the Kidds (that is, as an employee) and, therefore, was found not liable for trespass or payment of the damages. Had the contractor acted on his own, without direction from the Kidds, he might have had to share responsibility for the damage. The case of *Barnstead v. Ramsey* (1996) provides a contrasting view where the client and contractor were found to be sharing a common purpose and were found jointly liable.

Issues of joint liability and the role of independent contractors are discussed in more detail in Fridman (1990). In discussing agents and servants, Fridman noted that

> The ultimate question, in all cases, is whether what the employee did was sufficiently connected with what he was employed to do so as to lead to the logical, as well as practical conclusion that the employer should, and can be held responsible (Fridman 1990).

He goes on to note, in discussing independent contractors who cause damage incidentally in the course of their work and not as a result of prior authorisation, that

> ..it has been settled law that one employing another is not liable for his collateral negligence unless the relation of master and servant existed

> between them. Thus, a person employing a contractor to do work is not liable for the negligence of that contractor or his servants (Fridman 1990).

These issues relate directly to vicarious liability and are particularly important for the arborist who works as an independent contractor to understand. Some guidance on contracting and professionalism in arboriculture may be found in Chapter 10.

1.2.4 Damage and Damages

Once liability has been shown and accepted by the Court, the type and extent of damage and damages are assessed. Black (1990) defines the two terms as follows:

> Damage. Loss, injury, or deterioration, caused by the negligence, design, or accident of one person to another, in respect of the latter's person or property. An injury produces a right in them who have suffered any damage by it to demand reparation of such damage from the authors of the injury. By damage we understand every loss or diminution of what is man's own, occasioned by the fault of another. The harm, detriment, or loss sustained by reason of an injury.

> Damages. A pecuniary compensation or indemnity, which may be recovered in the courts by any person who has suffered loss, detriment, or injury, whether to his person, property, or rights, through the unlawful act or omission of another. Money compensation sought or awarded as a remedy for a breach of contract or for tortious acts.

So, the actual effects are the damage (injured person, damaged house, loss of screening and so on). The dollar values attached to this physical damage are the damages. Damages are assessed in many ways. Common forms in tree related cases include:

- **Actual or Compensatory Damages**
  The amount of money awarded to compensate for the actual loss of value or the injury sustained (that is, the amount of money required to exactly offset the costs of the damage). These damages are designed to enable the person or property to be restored as closely as possible to the condition known prior to the incident. Two forms of actual or compensatory damages are recognized: general (or compensatory) and special damages.
  - General or Compensatory Damages
    Pecuniary losses that will occur in the future (loss of future earning capacity), and non-pecuniary losses resulting directly from the incident, such as damages for pain and suffering. General damages are typically difficult to assess accurately.
  - Special Damages
    Pecuniary losses incurred up to the date of trial. They can be accurately assessed and defined. For example, if a falling tree hits a house, the cost of repairs to the house are shown in the invoices and bills. Special damages are normally specifically argued and proven in court.
- **Consequential Damages**
  The damage, injury, or loss of values that occurred due to other direct but perhaps delayed consequences of the act or incident. This damage, injury or loss of values might not have been ordinarily foreseeable.

- **Nominal Damages**
  Slight or very minimal damages sufficient to require token compensation.
- **Exemplary or Punitive Damages**
  Damages greater than actual damages; awarded because the incident resulted from malice, fraud, or wilful and wanton disregard by the defendant. Such damages help to compensate for the unnecessary aggravation caused by the incident and are designed to punish the defendant for his or her actions, as well as to set an example to others.
- **Aggravated Damages**
  Damages designed to compensate the injured feelings of the plaintiff (as opposed to penalising the defendant) when the injury results from malice or outrageous conduct. Typically seen in intentional tort actions, but can also arise in negligence cases when the plaintiff's trauma is exacerbated by the actions of the defendant.
- **Criminal Damages**
  Normally assessed for acts other than fire or explosions, criminal damage includes the wilful injury, damage, mutilation, defacement, destruction, or substantial impairment of the use of any property belonging to or enjoyed by someone else, without that person's consent.

Once the extent and conditions of damages have been determined, they form part of the final judgement. If the award is deemed by either party to be wrong in fact or law, an appeal may be launched. It is difficult to change a finding of fact by the trial judge, and most appeals are based on errors of law. In a few cases the provincial court of appeal decisions have been appealed to the Supreme Court of Canada. The latter then rules on the whole case, and often sets new precedents on the finer points of the law. The Supreme Court may also order a new trial. In researching legal principles, the precedents set by the supreme court cases (both provincial and national) carry the greatest weight.

1.3 Third Party Actions

Not all court cases involve just the plaintiff and the defendant. Sometimes a third party action is initiated by the defendants in order to try to include others who may also be liable for the damage. Third party actions are distinct from the main court action between the plaintiff and the defendant. In a third party complaint, the third party may be found liable for all or part of the damages which the plaintiff may win against the defendant.

For example, in *Timpauer v. Coffey* (1991), two third party actions were launched in addition to the main case of tresspass. In this case, the plaintiffs, the Timpauers, and the defendants, Mr. and Mrs. Coffey, were neighbours. The plaintiffs purchased their home because of the park-like setting. Unfortunately, during the clearing work for the defendant's new home next door, trees contributing to the park-like atmosphere and located on the plaintiff's property were removed without the plaintiff's consent. The Timpauers launched an action against the Coffeys and Anthony Wood Contracting Ltd. The judge awarded compensatory damages of

$31,400 against both defendants and punitive damages of $7,500 against the Coffeys alone.

The Coffeys launched a third party complaint against Anthony Wood Contracting Ltd., alleging that the contractor was at fault. The Court dismissed this action. A second third party complaint was filed by Anthony Wood Contracting Ltd. against the Coffeys, alleging that the tree clearing contractor was working solely at the request of Mr. Coffey and therefore should be indemnified by the Court in this action of trespass. The Court agreed, and ruled that the tree clearing contractor be compensated by the defendants, Mr. and Mrs. Coffey, for the amount awarded to the plaintiffs in the main action. In total then, the Coffeys ended up with a bill for $38,900 plus court costs.

In a third party action in *Kates v. Hall* (1990), the defendant, Hall, applied for indemnity under his insurance policy so that the insurance company would pay the damages awarded by the Court and the costs of his defence. Since intentional property damage was specifically excluded from the defendant's policy and he did not comply with the requirements of the policy, the insurance application was denied.

1.4 Conflicting Evidence

Conflicting evidence is quite commonly presented. The judge (and in some cases the jury assisted by the judge) has to determine the facts of the case. Judge and/or jury then becomes the trier of fact—they alone have to decide who is believable and which facts are acceptable. See Chapter 10 for some discussion of how conflicting evidence can affect consulting arborists.

There are many examples of conflicting evidence, of which *Hik v. Kerr* (1992) is typical. The expert witness for the plaintiff did not count the trees that were cut but concluded that 58 trees were felled, based on a comparison of the size of cleared areas with the areas which still had trees. The clearing equipment operator said he removed 30-35 trees. The defendant called an expert witness who used photogrammetry to estimate the number of missing trees. The photogrammetrist obtained aerial photos taken before the area was cleared and other photos taken after the cutting took place. He compared the photos and concluded that 14 trees were missing. The photogrammetrist admitted that he made some errors preparing the charts and that he could have missed a number of trees in the before photos because of shadows. He admitted that more than 14 trees were missing but could not be specific. The Court accepted the evidence of the equipment operator and concluded that 35 trees were missing, since this number was more consistent than the evidence of the photogrammetrist.

There are many examples of conflicting evidence...

In *698419 Ontario Ltd. v. Biggs* (1994), the condition of a Scots pine was disputed. The tree, according to the defendant, was almost dead. He alleged that the tree had been damaged by ice and had a longitudinal crack along the trunk. He said that there was no growth on the main trunk and the tree had a 90 degree turn in its top two metres. The plaintiff described the tree as straight and in reasonable condition with no crack in its trunk. He said that the tree had most of its limbs but the top was growing at a right angle to the soil. The Court rejected the evidence of the defendant and ruled that it was a typical Scots pine.

In *Sinkewicz v. Schmidt* (1994), aerial photographs provided evidence which contradicted the tree count alleged by the plaintiff. Six eyewitnesses also disputed the plaintiff's claim that 200 mature trees had been growing on the site. One eye witness estimated that there were only six mature maples on the site. In addition, the plaintiff claimed that he had planted, watered and weeded 1,690 new trees and shrubs. Five witnesses testified that they saw no newly planted trees on the land nor the plaintiff tending to them. Such discrepancies cast doubt on the credibility of the plaintiff.

In many cases, neither the plaintiff nor the defendant had any idea of the precise location of the property line or how many trees were actually cut or removed. Ambiguities in the facts of these court actions have commonly lead to counter claims by the defendant. Finally, while parties may call expert evidence, the Court is not obliged to accept any given set of evidence in its entirety. For example, in *Kates v. Hall* (1990), the Court did not accept tree replacement costs set out by either party.

1.5 Summary

Owning one or more trees or being assigned responsibility to look after trees entails a duty of care. The standard of care, which is the extent to which the duty of care must be exercised, is determined by what a prudent or reasonable person ought to have known, or ought to have undertaken. In the event that this duty of care is breached, there may be a case of negligence, and ensuing liability for the resulting damage. The extent of the award made, termed the damages, is determined by the degree to which a reasonable standard of care has been breached, the type of damage caused, and the degree to which the damage was intentional or wilful. In some instances, one or more third parties may be included in the court action. In most cases, the evidence presented by either side will be in conflict, and the judge and/or jury have to decide which evidence is factual in order to render a decision.

In the following chapters, specific types of tree related court cases are examined in more detail, along with the relevant precedents.

Chapter 2
Valuing the Damage
2.1 Introduction

Valuing the damage involves determining the type of damages (defined in Section 1.2.4) to be awarded and then assessing a reasonable dollar value for those damages. In discussing the types of damages, particular attention is paid to general or compensatory and exemplary or punitive damages because they are most commonly assessed by the courts. Consequential, aggravated, or criminal damages usually are awarded in conjunction with general (or compensatory) or exemplary (or punitive) damages and so are mentioned in that context. Nominal damages are discussed only incidentally because they are awarded where the injury is slight or minimal and was unintentional.

2.2 Determining the Type of Damages

2.2.1 General or compensatory Damages

General or compensatory damages are commonly granted by the Court in cases where tree injury is substantial. The idea behind general or compensatory damages is to restore the plaintiff's land to a condition reasonably close to that which existed prior to the damage. General or compensatory damages have been awarded for losses of land value, income, and amenities such as privacy, aesthetic value, shelter from the prevailing winds, or surface water run-off reduction. Other costs, such as those of cleaning up, replacing trees, and restoring the land to its original condition, have also been considered general or compensatory damages. The successful party in a lawsuit will often receive compensation for legal expenses, expert witness costs, special costs, and/or even prejudgement interest on pecuniary losses.

In determining the proper amount of damages to be paid, the wishes of the plaintiff are subordinate to the overall consideration of what is reasonable, practical, and fair. In awarding general or compensatory damages, the Court is concerned with reinstatement as opposed to retribution. The Court must weigh all of the evidence and apply common sense, reasonableness, and fairness among all parties in light of all the circumstances (*Boudreau v. T.P.Downey & Sons Ltd.* [1987], *Kates v. Hall* [1989], and *O'Cadlaigh v. Madiuk* [1994]). The trial judge in *Richardson v. Avery* (1995), rejected the plaintiff's claim for $250,000 as "quite outrageous" when 20 second growth trees were cut and harvested from the plaintiff's property. The Court awarded a more reasonable figure of $6,858.88 for damage repair and for the value of the wood harvested. In *Boudreau* the trial Court awarded the plaintiffs $7,000.00 in general damages. This award was subsequently appealed and overturned, with the final award being reduced to $250.00 in general damages.

...the wishes of the plaintiff are subordinate to the overall consideration of what is reasonable...

Loss of amenities may include loss of enjoyment of the land due to a loss of privacy or of the aesthetic value of the trees. An amenity contributes to the pleasantness or desirability of a property as a residence or to the pleasure of the residents rather than to their fundamental needs. In the cases cited below, awards were made for loss of amenities.

In *Crawford v. Crawford* (1986), trees were removed from the rear centre portion of the plaintiff's property. The removal of the trees reduced the privacy surrounding the plaintiff's swimming pool. The Court assessed general damages at $1,500 and ruled that the plaintiff was entitled to more than nominal damages for trespass. The damages were not confined to the value of the trees, which could not be replaced, but included an additional amount to compensate the plaintiffs for the loss of privacy to their property and the loss of the aesthetic value of the trees destroyed.

The plaintiffs in *Kates* (1989) were awarded loss of amenities for the reduction of privacy around their house. In *Hamel v. Moilliet* (1994), the defendant encroached 26 metres onto the plaintiff's land and cut down 15 large trees. Some of the trees were felled onto adjacent trees, breaking their tops off and causing other damage. When the defendant skidded the logs out, the skid road damaged the underbrush and more trees. The plaintiff found that the absolute privacy she had enjoyed was gone. The Court awarded the plaintiff $20,000 for general damages for the trespass and loss of amenities, \$6,300 to replant 3.7 metre to 4.6 metre tall fir trees, plus punitive damages of \$7,500, for a total of \$33,860.

In *Horseshoe Bay Retirement Society v. S.I.F. Development Corp.* (1990), the plaintiffs owned four residential properties known as Lots 35, 36, 37, and 38. Each of the lots was 15 metres wide and ranged from 61 metres to 70 metres deep. The centre and rear of Lots 35 to 38 were, before the trespass took place, heavily wooded with a second growth forest. The damage to Lot 35 was relatively slight. In Lot 36, well over half of the trees were removed. Three Douglas-fir, three bigleaf maple and one western red cedar were felled. Six of the seven trees were left with three to six metre high stumps. The judge observed that "Whereas the area was once wooded, it now looks like a bunch of abandoned telephone poles." Approximately 75% of the trees on Lot 37 and 50% of the trees Lot 38 were cut.

In this case, the owner of Lot 35 received \$922 for the loss of her trees and \$300 to clean up the logging debris. The owners of Lot 36 were awarded \$9,527 for their loss of privacy, loss of shelter from the prevailing winds, and loss of control of surface water run-off, in addition to \$3,000 for clean up. The owner of Lot 37 was awarded \$6,511 for the loss of the aesthetic value of her trees, \$4,000 for damages resulting from flooding due to the removal of the trees on her property and to offset the risk of further flooding to her home, and \$2,000 for clean up. An award of \$5,112 was made to the plaintiff who owned Lot 38 for loss of privacy, loss of shelter from prevailing winds, and loss of control of run-off, in addition to $1,800 for clean up.

In *Timpauer v. Coffey* (1991), an award of $10,000 was made to the plaintiffs for loss of amenities. In this case, the park-like setting of the plaintiff's property was due to the dense tree growth. When the trees were unnecessarily cut down, the park-like amenity was lost because the replacement trees would take many years to reach a similar height to the trees removed.

In *Muncaster v. Nunnemacher* (1994), a contractor employed by the defendant cut 13 mature trees on the land of the plaintiff. An award for amenities was made because the trees had formed a screen between the plaintiff's and defendant's lands. The view of mature trees was replaced by a high wall surrounding the defendant's new swimming pool. The defendant was concerned about pine needles dropping into the pool. The plaintiffs, the Muncasters, claimed that they had to change the design and construction of their house as the result of the removal of the trees. They claimed general damages for trespass and for nuisance or, alternatively, for negligence. The plaintiffs also claimed $100,000 in punitive damages. They were awarded $13,300 in compensatory damages for the loss of amenities (their view of the trees) and $5,000 for punitive damages. They appealed the award but the appeal was dismissed with costs (*Muncaster v. Nunnemacher* [1996])

In an example from public lands, *Prince Rupert (City) v. Pederson* (1992), the defendant cut down and removed 89 trees (71 hemlocks and 18 western red cedars) on city park land. The trees ranged from 12 to 15 metres in height. The Court ruled that a loss of amenities to the public had occurred. In addition, the trees provided wildlife habitat, a play area for neighbourhood children, and a screen against wind and noise. The park land was now subject to erosion and the trees on the edge of the cut line were blowing down. The Court ruled that complete restoration of 89 trees of similar height was not reasonable. A partial restoration with 3 to 3.6 metre high conifers, including the cost to supply the trees, prepare the site, and install the trees, for a total of $63,365 was considered reasonable. The Court also assessed the general loss of public amenities at $10,000, since it would take some time for the smaller trees to revegetate the park land.

The plaintiff in *O'Brien v. Underwood McLellan & Associates Ltd.* (1979) was the owner-operator of a lakeside tourist facility, a hunting and fishing camp on Green Lake in Saskatchewan. The defendant, while conducting a feasibility study for the construction of an air strip, trespassed onto the plaintiff's property and cut a 122 metre by 30.5 metre wide swath through the plaintiff's trees. In addition to the trees damaged, the soil on the plaintiff's land began to erode. Eventually the plaintiff subdivided his property and sold the land as lakeside lots. He was awarded damages for the difference in the value of the four lots before and after the defendant cleared the trees.

Loss of Income

In several cases the courts have awarded loss of income. In *Howard v. Morgan Trust* (1983), the plaintiff missed 13 working days as the result of the stress caused by the cutting of the trees and the work involved in making an inventory of the trees felled. The Court awarded $500 for the inconvenience and loss of salary.

In *Lee v. Hersh* (1993), the plaintiff was a gynecologist and obstetrician practicing in Vancouver, BC. Dr. Lee maintained that, as a result of the trespass and resulting damage to the trees, driveway, sidewalk, and garage, she was required to pay a substitute physician to take her place to allow time to deal with the damage to her property. She received an award of $16,000, which was substantially less than the amount she claimed, to compensate her for the loss of income and the amount required to pay the substitute. In addition Dr. Lee received $36,556.20 in special damages to cover the cost of repairs to her property and $22,400 to replace the trees damaged. The defendant was also assessed $23,000 for punitive damages. The judge concluded that the defendant's behaviour was "indifferent, arrogant, intentional, outrageous and highhanded."

Emotional Stress

The plaintiff in *Goodwillie v. Frank* (1991), claimed damages for emotional stress as the result of trespass and damage to trees on his property. The Court acknowledged that the plaintiff was very annoyed and unhappy about the tree pruning but suggested that the plaintiff overreacted. An award for emotional distress, particularly since there was no medical evidence to substantiate it, was not granted by the Court. In *Boudreau,* the plaintiff claimed damages for anxiety, mental anguish, upset, and emotional distress. Again, the Court rejected the claim.

Other Costs

Various costs, such as those of cleaning up, replacing the trees, and restoring the land, may be considered types of general or compensatory damages. All four plaintiffs in *Horseshoe Bay Retirement Society* were awarded the cost to clear away the logging debris on their lands. In *Howard,* the plaintiff was awarded the cost to remove the felled trees. In *O'Cadlaigh v. Madiuk* (1994), the defendant was assessed $4,800 by the Supreme Court of British Columbia for clean up costs. Defendants have an obligation to clear away the mess resulting from tree cutting.

Many courts award the cost to replace the trees damaged or cut (*Cosco v. Scott* [1984], *Howard*, *Lee*, *Peters v. Diamond* [1964], and *Timpauer*). The Court may also award the cost of repairs or the cost of restoring the land back to its former state (*Kates* [1989], *Lee*, and *Prince Rupert [City]*). One of the important points to note is that the award for replacement must be deemed by the Court to be reasonable (*Kates* [1989] and *Prince Rupert [City]*).

The cost to hire an expert is often claimed. In *Howard*, for example, the plaintiff was awarded $4,731.33 to pay for the services of the expert witness.

Prejudgement interest is often awarded on pecuniary losses such as special damages and loss of income, and is calculated from the actual date of the trespass until the judgement date. Usually there is no prejudgement interest assessed on non-pecuniary losses, general or exemplary damages, the size of which are determined at the time of the judgement.

Special court costs are usually awarded when the party or his or her lawyer has behaved improperly and wasted the Court's time. For example, special court costs were awarded in *Prince Rupert (City)*, where the Court found that two days had been wasted by a parade of witnesses whose testimony was irrelevant to the main issue of the action. The Court awarded the plaintiff special costs for two days of trial. Other examples in this category include: *Chan v. Rowley* (1992), *Anderson v. Skender* (1991), *Michaels v. Zwickel* (1994), and *Stowe v. McTaggart* (1993).

2.2.2 Exemplary or Punitive Damages

Exemplary or punitive damages have been awarded in addition to general or compensatory damages. As discussed in Section 1.2.4, an award of punitive damages means that the defendant has, with knowledge of the consequences, willfully damaged the trees and plants, often with callous or indifferent disregard for the property rights of the plaintiff. The idea behind punitive damages is to punish the defendant or to set an example to others.

In *Horseshoe Bay Retirement Society*, a neighbour complained about the tree cutting because the defendants were felling trees beyond the property line. In response, the defendants ordered the workers to stop the tree cutting, but two weeks later they returned with four fallers and resumed logging the trees. The defendants deliberately entered neighbouring lands and cut a swath through the forest to create a view. In assessing punitive damages, the Court ruled that the increased selling price of the houses built on the new view lots should be taken into account. Since the trespass and cutting of the trees was deliberate and malicious, the Court suggested that other developers who might consider similar acts should be warned that this type of activity was unacceptable. Exemplary or punitive damages were assessed at $100,000, substantially more than the increase in the selling price of the of the defendant's new view lots.

In his judgement on this case, Justice Shaw referred to principles for awarding exemplary and punitive damages which had been identified in the earlier cases of *Austin v. Rescon Construction (1984) Ltd.* (1989), *Merest v. Harvey* (1814), and *Pretu v. Donald Tidy Co. Ltd.* (1966). These were:

1. Exemplary or punitive damages may be awarded where the defendant's conduct was malicious or high-handed.

2. That in assessing the damages the degree of culpability of the defendant's conduct should be taken into account.

3. That if only a nominal amount were added to the actual damages, this would be but an invitation to contractors to violate property rights, for what would amount to an

insignificant licence fee and that a substantial amount for punitive damages should be added to the actual damage suffered by the plaintiff.

4. ... in a case where a man disregards every principle which actuates the conduct of a gentleman, what is to restrain him except large damages?

5. ... the damages should be sufficient so they will be a deterrent and all will know that it will not be profitable to trespass upon the lands of another, and the amount to be added to the damages for the loss suffered by the plaintiffs should be substantial.

In *Prince Rupert (City)*, the value of Pederson's land increased $14,000 as the result of the removal of trees on the park land adjacent to his property. Pederson's actions throughout were deliberate and in contempt of the rights of the public. The Court suggested that Pederson's contention that he honestly mistook where the park lands lay "was really like a small boy caught stealing jam." The Court determined that an indiscriminate cutting of trees occurred on the park land for the personal improvement of Pederson's property. The award for exemplary or punitive damages in this case was $58,500 based on the $14,000 increased value in his property and $500 per tree for the 89 trees cut and removed. While the courts may not deem it appropriate to assess general damages in the amount of the financial benefit gained by the defendant's action, as in this case, such benefit may be considered when awarding exemplary or punitive damages.

The defendant, Pederson, appealed the judgement (*Prince Rupert [City] v. Pederson*

[1994]). The appeal was allowed in part when the court reduced the award of general damages by $10,000. The court ruled that the City of Prince Rupert, a municipal corporation and therefore an inanimate body, could not suffer a loss of amenities. In addition, the city had no status to sue on behalf of its residents for their loss of pleasure, protection, and shade. The award of punitive damages was also reduced from $58,500 to $25,000. The judge reduced the award because the value of the land for which Pederson was attempting to get a view was $25,000. At the time of the trial, Pederson had not sold the property and had not gained a profit from the trespass. The sought after view would disappear soon after the area was replanted. Court of Appeal Judge J.A. Southin stated:

...trespass does not pay

> While it is not erroneous in law to assess exemplary damages for tree cutting at so much a tree, the Court must then look at the figure so calculated and ask whether it is too much. Exemplary damages are a form of punishment and the criminal law sentencing principle of totality is fairly applied by analogy.

Judge Southin remarked that greed was Pederson's sole motivation in cutting down the trees and that this case should provide a reminder to aspiring trespassers that trespass does not pay.

In *O'Cadlaigh,* the Court found that the defendant pointed out specific trees which he instructed the tree clearing contractor to cut. The defendant knew that the trees indicated were on the plaintiff's land. The Court awarded

$25,000 in exemplary or punitive damages to the plaintiff.

Often at issue is whether the tree cutter had permission, express or implied, from the plaintiff to cut and remove the trees. Sometimes defendants admit that they did not have permission to cut trees. In *Cordiez v. Muffly* (1991), the defendant said in his discovery statement that he was treating the land of the plaintiff as his own land and keeping it clean by removing the trees. He thought there was a working agreement to that effect, but clearly he was mistaken. The Court concluded that there was no tacit right or permission given to Mr. Muffly to cut trees or to otherwise enter onto the plaintiff's land. The Court, in response to Mr. Muffly's high-handed conduct in cutting, burning and girdling trees on his neighbour's lands with the sole purpose of improving his view, awarded the plaintiff $5,000 for exemplary damages. Similar issues arise in *Anderson*.

2.3 Assessing the Cost of Damage

The exact manner in which a dollar value for damage is determined may be a matter of debate. The issue before the courts boils down to whether damages should be measured in terms of the reduced value of the property or in terms of the cost to repair the property to its original condition. Sometimes, however, real estate appraisals may cite no loss or even an increase in property value when trees are removed. In such cases, the amount needed to repair the damage may be substantially greater than the amount by which the property has been devalued. Since there is judicial authority for both approaches, the Court decides between them by attempting to test the reasonableness of the plaintiff's desire for restoration of the land to its former state. The trees themselves only need to be valued if the second approach is chosen.

2.3.1 Reduction in Real Estate Value

In *Sinkewicz v. Schmidt* (1994), the defendant, a municipality, bulldozed the plaintiff's trees to fill an excavation on an adjacent property. The plaintiffs, in this case, claimed for the cost of reinstating their land to its original state. The defendant claimed that the damages should be assessed by determining the reduction in the property value due to the loss of the trees. The Court weighed the advantages to the plaintiffs of replacing the trees as opposed to the added cost to the defendant of paying for the trees rather than paying for damages for the reduction in property value. As part of this test of reasonableness the Court asked how the plaintiff had used and would use the land. In this case, the plaintiffs were awarded the depreciated selling value of their property. The land was worth $4,900 before the trespass and $1,500 after; the Court awarded the difference of $3,400 as damages.

In *Scarborough (Borough) v. R.E.F. Homes Ltd.* (1979), the defendant removed three silver maples which were growing on a municipal road allowance. The Court used the diminution in the value of the land, as the result of tree removal, to assess damages. It noted that it was difficult to prove diminution in the value of a road allowance due to tree removal, because road allowances are not bought and sold on an open market. Therefore, the cost of restoring the land to its former state was used to calculate the diminution of land value. The plaintiff's expert witness employed the CTLA formula method to value the three trees at $12,069. Both the trial judge and the appeal court judge rejected this method of valuing the trees, saying that the diminution of value of a road allowance was a loss, but different from the loss of trees on residential, commercial, or industrial lands. At the judge's discretion, an intermediate value of $4,000 was awarded to the plaintiff.

In *O'Brien,* the plaintiff was awarded damages for the difference in the value of the lots before and after the defendant cleared the trees for an air strip.

2.3.2 Valuing the Trees

Trees may be valued in several ways, including: estimating the costs of restoring the damaged land, known as a cost of cure approach; applying the trunk formula method promulgated by the Council of Tree and Landscape Appraisers (CTLA); and assessing the timber value of the logs cut.

Cost of Restoration

In *Milliken v. Kidd and Kidd* (1994), a contractor hired by the Kidds unwittingly trespassed and cut down nine eastern white cedar trees and one maple. The judge examined the costs of restoring the site to its predisturbance condition, the depreciation of property value, and replacement costs. He concluded that, based on other similar cases in Ontario, the cost of replacement trees could be used to determine the damages. The judge reviewed several quotes for replacement costs and awarded the Millikens $6,000 in damages, plus $200 for time and inconvenience.

There are a number of cases where the damages awarded reflect the cost of restoring the land to its former state

There are a number of other cases where the damages awarded reflect the cost of restoring the land to its former state. Examples include: *Carr-Harris v. Schacter* (1956), *Dykhuizen v. Saanich (District)* (1989), *Horseshoe Bay Retirement Society, Kates* (1989), and *Peters v. Diamond* (1964). In *Peters* the judge states:

> I do, however, find that the defendants must place the plaintiff as nearly as possible in the same position as he was before he sustained the damage by the defendant's tort. This can only be done by assessing the probable cost of planting similar trees of as near a size as possible and practicable, and such as will afford to the plaintiff a similar amount of screening, privacy, beauty and of attraction to his children. I can not add anything to this for depreciation in market-price of the land.

The cost of replacing damaged trees is calculated using the nursery retail cost of plant material plus the cost to replant and restore the area to its former state. In *Chan v. Rowley* (1992), the Court rejected the conflicting tree values provided by experts from both sides. Instead, the Court chose a value of $4,000, which represented the projected cost of substituting 3.7 to 4.3 metre specimens for the three large cedars that had been cut down. In the Court's opinion, this replanting would not replace the original mass of the trees but would over time provide a screen.

Other examples in which the courts accept a replacement value to compensate for damage and loss of trees include *Cosco v. Scott* (1984), *Barnstead v. Ramsey* (1996), and *Kates* (1989). In *Cosco,* the award was based on the sum of a landscape contractor's estimate of the costs: to replace the trees, to remove the stumps and further clean up the site, to ship the trees, and for the labour to install them on site. In *Barnstead,* the Court accepted the restorative measures put forward by a landscaping company and awarded the plaintiff $1,700 for the cost of repair and $5,000 for general damages, including an assessment for lost screening and other amenities.

In *Kates (1989),* the proposal to replace the 13 hemlock with evergreens 12 metres in height was summarized by the trial judge of the B.C. Supreme Court in this way:

> They have obtained quotations of $195,000 and $210,000 as the price of bringing 12 metre replacement hemlock trees from the Fraser Valley and planting them on the property. A contractor who transplanted similarly sized trees from the valley onto the Expo site in downtown Vancouver says that with 18 to 24 months for preparatory work, and the mobilization of heavy equipment including two 65-ton mobile cranes, the job could be done. It would take three or four weeks of quite destructive work on the Kates' property, but the contractor says this could be remedied thereafter. The trees would have to be supported for at least two years with

guy-wires and anchors, and they would come with no guarantee of ultimate survival.

In addition to the disturbance to the Kates' land, several of the guy-wires would have to be anchored on the neighbouring property. The judgement on this case was rendered in 1990 and appealed in 1991.

The Appeal Court judges for *Kates v. Hall* (1991) concluded that this undertaking was massive. They dismissed the appeal and upheld the ruling of the lower court that the plaintiffs were not entitled to an average sum in excess of $200,000 because it was seen as unreasonable under the law. In *Kates* (1990), it had been noted that there was no measurable market reduction in the value of the property as the result of the tree cutting. A loss of privacy had occurred but was limited. Nor had the tree removal changed the character of the overall appearance of the property.

Madame Justice Proudfoot, in *Kates* (1991), referred to the case of *Dykhuizen v. Saanich (District)* (1989),

> The law in British Columbia following *Dykhuizen* requires the Court to ask what is reasonable in the circumstances, not… …what are the "express wishes" of the appellants. As the trial judge said [in *Dykhuizen*], "The damages in such cases may extend to the cost of restoration or restitution within reasonable bounds, together with compensation for loss of amenity to the extent that complete restoration cannot reasonably be affected." and "The authorities clearly establish that there are limits imposed on the extent to which a trespasser who causes damage can be required to restore the land to its previous state. A court will not

> require the trespasser to finance restorative measures which are plainly unreasonable."

Therefore, the Court refused the claim for damages which involved the excessively expensive transplanting of very large replacement trees, ruling that it was plainly unreasonable. The BC Court of Appeal upheld the lower court's award to the plaintiff for compensatory damages of $21,000 for reasonable remedial work and $1,000 per tree for loss of amenities while the replacement trees grew to maturity. The Court also found the defendants' conduct in cutting the trees high-handed and awarded exemplary or punitive damages of $2,000 per tree. The Appeal Court, in upholding this award, found that, in fact, it could have been assessed higher (*Kates* [1991]).

When assessing the plaintiff's damages, the Court in *Hamel v. Moilliet* (1994) used the sum of $1,000 per tree for large trees as approved by the Court of Appeal in *Kates* (1991). In addition to the $15,000 allowed for the 15 trees cut, the Court increased the award by $5,000 to allow for the damage to the undergrowth which was caused by the skid road.

In *Cordiez*, the parties had earlier agreed to a value of $50 per tree as appropriate compensation. The Court in the absence of any other or better evidence, fixed damages for the loss of the 28 trees at $1,400 plus $300 to compensate for the slash burning around the trees, for a total of $1,700 in general or compensatory damages.

In some instances the values claimed by the two parties differ widely. In *Oran v. Westwood*

Fibre Ltd. (1996), the plaintiffs claimed damages of $148,000, comprising $50,000 for the trees removed, $20,000 for damage to the undergrowth, and $78,000 for loss of amenities. The defendants admitted the trespass, but claimed it was entirely accidental. The judge noted that the area affected was about ¾ of an acre in a 5.5 acre parcel and had not previously been landscaped. Given the plaintiffs intention to not occupy the land until 2004 (their planned retirement date), the replanted trees would have grown back to a height sufficient to restore privacy and aesthetics. An important point was noted in this case:

In some instances the values claimed by the two parties differ widely

> The award for cost of restoration must relate to the reduction in value of the property flowing from the defendant's conduct. That is, one of the factors to be considered in determining what are reasonable damages is the cost of restoration of property in proportion to any diminution in the value of the property. In determining what is reasonable restoration, the Court must also be mindful of the uses to which the plaintiff has and will put the property and the Court should, where the plaintiffs have taken no steps to restore the damage caused by the trespass, weigh the lack of that restoration against evidence of the plaintiff regarding the value of the trees and claims for loss of amenities.

Citing *Kates* (1991), the judge then awarded the plaintiffs a total of $28,500 plus their costs, this being restoration at $20,000, loss of amenity at $3,500, and cleanup at $5,000.

The CTLA method (also known as the ISA trunk formula method) uses the cross sectional diameter of the tree trunk multiplied by a base dollar value (whose method of determination may vary slightly by region) to yield the basic value of the tree. This value is then adjusted by a series of factors to allow for species, condition, and location (location being a combination of site, contribution to the landscape, and placement within it). The end figure can then be used as a reasonably objective test of tree value. It must be noted, however, that the trunk formula approach was specifically designed for trees (naturally arising or planted) that have demonstrable amenity value. It is not always applicable in cases where the trees form part of a woodlot or forested area. In some cases, a combination of approaches has been adopted (for example *Sanda v. Caple* [1987]).

In *Dykhuizen v. Saanich (District)* (1989), the CTLA trunk formula method was used by the District of Saanich to establish the value of three western dogwoods, two bigleaf maples and an alder. The horticulturist employed by the municipality said that this value was much lower than the replacement cost for similarly sized trees. Nevertheless, the County Court of Vancouver Island accepted the value given by the CTLA trunk formula method and awarded the District of Saanich $6,371 in compensatory damages and $1,000 in exemplary or punitive damages. The award was appealed in the British Columbia Supreme Court.

The appellant's counsel contended that the tree valuation method put forward by the municipality and adopted by the lower court did not provide a reasonable basis for the assessment of damages to native trees. The appellant's counsel contended that the municipality suffered no damages because the trees would have had to be removed anyway when the municipality approved the subdivision. The Supreme Court ruled that this view was not valid, since the action was for deliberate or willful trespass. A willful trespasser was not entitled to say that the damage he or she did would probably occur in any event and should for that reason be excused. Moreover, the Supreme Court ruled that the CTLA formula used for assessment of general damages was reasonable, and that the defendant proceeded with sufficient disregard for the property rights of the municipality to justify the exemplary award. The appeal was dismissed.

A willful trespasser was not entitled to say that the damage he or she did would probably occur in any event and should for that reason be excused

In *Howard v. Morgan Trust* (1983), the expert for the plaintiff used the method established by the CTLA to assess the damage for the 546 stems cut by Timmerlinn Ltd. at \$18,151.72. The defendants' expert allowed only the wood value of \$394.97, plus an assessment of \$3,016.05 for two white and one grey birch which were classified as shade trees located close to the housing site. Mr. Justice Gonthier, in his judgement in the *Howard* case, stated:

> ...in this context the ISA method is the only one put forward in evidence to evaluate a loss of amenity. Indeed this method expressly provides for the evaluation amongst others of unmanaged or dense forest area. The Court does not see why the enjoyment of a landscape may not equally rest in an area in its natural state as in an area which had been worked upon by man.

The judge further observed that

> ...both expert witnesses have explained the ISA method is but a method and approach, a guideline to the evaluation of damages and the percentages which it sets are not absolute, but merely guides which require the exercise of judgement in their application.

In assessing the damages in *Howard*, the Court sought to fully compensate for the loss suffered by the plaintiff but at the same time only to award those damages which had been proven. Therefore, the Court lowered the evaluation amount prepared by the plaintiffs' expert by using of a reduced location factor in the CTLA formula.

In *Anderson v Skender* (1991), the Court accepted the CTLA formula method but reduced the amount calculated by the arborist by 25% to arrive at an award for the value of the trees. The deduction was made in favour of the Skenders on the basis of a 25% ownership interest in the double trunk tree that was damaged but was growing entirely on the Anderson's property. This method of using the percentage of the tree on either side of the boundary line to apportion damages fairly between parties was also employed in *Kelley v. Ingersoll (Town)* (1996), discussed in Section 8.4. In assessing the amount for the punitive damages in *Anderson*, Mr. Justice McKenzie discussed the CTLA formula method. He considered the value placed on the trees by the expert witness for the plaintiff when he said:

> *The Guide for Judging the Condition of Landscape Trees* developed by the Arboriculture Society and applied by Mr. Bottrell awards separate points for crown development, trunk condition, major branch structure, twig growth rate, foliage, insects and diseases and roots. It considers position and other factors to determine location values and the shade factor. It is very detailed and has many objective features but, inevitably, some subjective factors such as judging the aesthetic quality of any tree. Mr. Bottrell belongs to the Arboriculture Society and is a proponent of the values they espouse which include the preservation of trees in an urban atmosphere. I discern in his evidence that he and the Society have a sense of mission about preserving trees and are opposed to their willful destruction by developers. I believe this not unworthy attitude had influenced upward to some extent the monetary values assigned by the formula to various trees. I believe that an indignation or retribution factor is built into the valuation formula. This factor has features in common with the principle of punitive damages so I will put some restraint upon the award for punitive damages.

Unfortunately, the amounts calculated by experts using the CTLA formula can differ considerably. In *O'Cadlaigh v. Madiuk* (1994), seven trees were cut, among them a mature western red cedar with a diameter at breast height of 117 centimetres and an approximate height of 24 to 30 metres. The plaintiff's expert, a certified arborist, valued the trees at $49,000. The expert called by the defendants, also a certified arborist, valued the trees at $11,720. An independent certified arborist inspected the large cedar shortly after it was felled. The evidence of this individual indicated that the tree was in poor condition, with decay evident. The tree also had a large crack up one side of the trunk which was exposed to the elements and the trunk was poorly shaped. The stump of the tree was sitting high on the surface of the soil and the roots were waterlogged well away from the stump itself. In addition the tree had co-dominant leaders and lacked structural integrity. Thus, the impartial certified arborist corroborated the facts observed by the defendant's expert. These facts were taken into consideration by the defendant's expert to arrive at his rating for the condition of the tree. He valued the cedar at $3,900. The plaintiff's expert valued the cedar at $26,000 because he gave the tree a higher condition rating.

Unfortunately, the amounts calculated by experts using the CTLA formula can differ considerably

In addition both the experts differed in their opinion as to the location value of the tree, with the plaintiff's expert giving the tree a high location factor. The impartial expert stated that the cedar was growing in a wilderness area. The cutting of the seven trees did not affect the plaintiff's privacy nor did it in any other way change the character of the site. The Court accepted the location rating for the cedar put forth by the defendant's expert. How did the Court choose? In this case the Court looked to the greater expertise and experience of the defendant's expert and also to the fact that the condition and location ratings used by this witness were independently corroborated.

In *Chan v. Rowley* (1992), the CTLA formula method was used to calculate the value of the three cedar trees removed. The plaintiff's expert valued the three trees at $53,593, while the expert witness for the defendant valued the same trees at $10, 328. The defendant's expert substantially discounted the value of the trees due to their location on the property. The Court looked at the lot value of $65,000 and rejected the $53,593 tree value as unreasonable since the removal of the trees did not materially change the character of the plaintiff's land. Also, there was no suggestion of any diminution in the value of the plaintiff's property. The Court also rejected the $10,328 put forward by the defendant's expert in favour of a value of $4,000 which represented the actual projected cost of replacing the three cedars with 3.7 to 4.3 metre specimens. The Court said that

> The opinions of experts are often used and provide useful assistance to the trier of fact. They do not go beyond that function. Apart from the methodology used by the experts in the circumstances of this case, there is always a subjectivity factor to assess because each of us may value trees differently.

Cost of Restoration and the CTLA Method

In *Lee v. Hersh* (1993), the value of the four elm trees cut down was calculated in two ways. An installed cost of $3,750 per tree was calculated by an experienced landscape contractor. The trees were also evaluated using the CTLA trunk formula method. The value per tree using this method was $2,740.95. The counsel for the plaintiff claimed $4,168 per tree installed. The Court awarded $4,000 per elm tree, plus $1,000 for each of the four trees for loss of amenities. Two additional trees were valued at $1,200 each for a total award of $22,400.

Timber Value

In *Harshenin v. Bayloff* (1991), the defendant trespassed and harvested timber on the lands of the plaintiff. In this case, the plaintiff was awarded an amount to compensate him for damage to his property and the value of the logs sold.

In *Bawa v. Noton* (1996), the defendant trespassed, cut timber from the land of the plaintiff, and sold the timber for a net profit of $8,379.81. The plaintiffs claimed for the net profit and for a sum of $5,000 to cover the costs of the clean up. In addition the plaintiffs claimed $49,211, which was the appraised value of the trees based on the CTLA trunk formula method. The Court rejected the third claim because the property was not residential but simply unimproved land. The judge also noted that the CTLA valuation method was not recommended for merchantable timber. The judge ruled that because the trespass was unintentional the plaintiffs were entitled to receive the value of the trees cut less the cost of logging and transport to market. If the trespass had been willful, the plaintiffs would have been awarded a greater amount.

2.4 Summary

The most common types of damages awarded by the courts are general or compensatory, and exemplary or punitive damages. Many cases have been cited where general or compensatory damages have been awarded for loss of amenity (privacy or enjoyment of the area). Punitive or exemplary damages have also been awarded in cases where it has been shown that the defendant acted with full knowledge of and wilful disregard for the effects of his or her actions. Further damages have been awarded for loss of income, the costs of hiring expert witnesses, and in at least one case, special costs due to the introduction of evidence that was deemed irrelevant to the case.

The dollar value for the damages may be assessed by determining the change in real estate value, the cost to to restore the land to an acceptable state, or the value of the trees lost or damaged. The cost of restoration and the CTLA trunk formula method are the most common approaches. The CTLA trunk formula method has not always been accepted because it incorporates a degree of subjectivity and is not applicable to naturally occurring or planted trees in woodlots or forests. The value of the trees as lumber has less often been used to assess damages. More frequently, damages are assessed by combining methods to ascertain the value of the trees damaged or lost plus the costs required to restore the damaged area. Damages awarded for repairing the land are seldom high. In most cases the award enables restoration of sufficient young tree cover to eventually compensate for the lost trees, rather than quick and complete replacement with mature trees, the costs of which are deemed unreasonable by the courts.

Chapter 3
Trees Damaged During Trespass

3.1 Introduction

Trees growing on public or private property have been damaged or destroyed for any number of reasons by trespassers acting unwittingly or with full knowledge of the consequences of their actions. In this chapter the cases reviewed focus on trees damaged within a property line or boundary. Trees growing right on a boundary line and tree branches that overhang a property line are discussed in Chapter 4. The basic legal aspects of trespass are outlined below.

3.2 What Is Trespass?

In many of the cases cited in this book, the issue hinged on who owned the affected trees and whether or not the defendant had a right to cross over his or her property line to deal with the tree(s) in question. In several cases, the defendants knew where the property line lay but deliberately went beyond it and cut or damaged trees outside their jurisdiction. In other cases, the trespass was unintentional, usually a result of either not bothering to check the exact location of a line, if it even existed, or of accepting someone else's word for the boundary location. The lack of a clearly and accurately defined boundary is a major cause of trespass and tree damage.

Black (1990) defines trespass as "The unauthorized intrusion or invasion of private premises or land of another." As with other legal terms, several forms are recognized, including:

- **Technical Trespass**
  The boundary line was crossed, but the damage caused was minimal and not wilful.
- **Inadvertent Trespass**
  The boundary line was crossed by mistake, usually because the location was genuinely believed to be at a different location.
- **Wrongful trespass**
  The boundary line was crossed because of indifference or negligence in determining where it was located.
- **Wilful trespass**
  The boundary line was crossed deliberately and damage inflicted in full knowledge of being beyond the boundary line.

The law of trespass is quite straightforward, and was referred to in *Webb v. Attewell* (1994) as follows:

> In an action for trespass, if the plaintiff proves the trespass he is entitled to recover nominal damages, even if he has not suffered any actual loss. If the trespass has caused the plaintiff actual damage, he is entitled to receive such amount as will compensate him for his loss. Where the defendant has made use of the plaintiff's land, the plaintiff is entitled to receive by way of damages such a sum as would reasonably be paid for that use. Where there is an oppressive, arbitrary or unconstitutional trespass by a government official or where the defendant cynically disregards the rights of the plaintiff in the land with the object of making a gain by his unlawful conduct, exemplary damages may be awarded. If the trespass is accompanied by aggravating circumstances

which do not allow an award of exemplary damages, the general damages may be increased.

The form of trespass affects the assessment of liability and damages. For example, if the Court finds that the trespass was accidental or inadvertent, damages may be more leniently assessed than if the trespass was wilful. In all cases, the Court will decide the basis upon which damages are to be assessed.

3.3 Types of Tree Damage

The types of tree damage that have come before the courts in Canada very widely. Most commonly, trees were cut down. Trees have also been topped and/or limbed, damaged by fire, or have had their trunks girdled. Many trees and shrubs have been damaged during construction activities.

In *Cordiez v. Muffly* (1991), 28 trees were cut down or damaged by fire. Several trees were girdled and left to die. The Supreme Court of British Columbia concluded that the defendant was in the habit of girdling trees in order to kill them while they remained standing, and assessed punitive damages as the result of this high handed conduct.

Six apple trees and one Manitoba maple tree were cut down by construction workers employed by the defendant in *Peters v. Diamond* (1964). In this case the fruit trees provided screening, privacy, improved the aesthetic quality of the property, and also provided a playground for the plaintiff's children.

Trees on neighbouring property have been topped or had their limbs trimmed back, as occurred in *Goodwillie v. Frank* (1991) and *Horseshoe Bay Retirement Society v. S.I.F. Development Corp.* (1990)

Sometimes trees are injured due to trespass during construction. In an example from Manitoba, *Petrillo v. Ikotin* (1993), trees and shrubs were damaged during the installation of a residential gas line. In *Nailen v. Powell River (Municipality)* (1978), a rockery, ornamental trees, and a set of steps were dug up and removed, and the property left in disrepair during the construction of a storm sewer adjacent to and partially on the land of the plaintiff. In *Carr-Harris v. Schacter* (1956), the defendant trespassed when he excavated a 3.7 metre deep hole to construct a wall for a garage on his land. His excavation extended under the plaintiff's garage, causing the soil along the property line to subside. A planting bed of shrubs and flowers and a 18 metre elm tree slid into the hole. The elm fell against another elm tree, damaging many of its branches and destroying its beauty. The first elm could not be replanted and was cut for firewood. In addition to the damage to the large trees, the boundary fence hung in mid-air over the excavation.

3.4 Reasons for Trespass

3.4.1 Accident

In some cases the defendant may have entered onto the property quite by accident. In *Mills v. MacAleese* (1987), the boundary line was wrongfully described by the owner of the land, and as a result, the tree clearing contractor trespassed and cut trees on eight hectares of neighbouring land. The plaintiff was awarded $10,200.00 in damages.

In *Barnstead v. Ramsey* (1996), trees were inadvertently removed when the defendants decided to cut down trees on their own property which were providing cover for burglars. The defendants hired a contracting company to cut the trees while they were away on vacation. The contractor trespassed and cut trees on an adjacent property by accident. The plaintiffs were awarded $5,000 in general damages and $1,700 (including GST) for restorative work.

...the defendant may have entered onto the property quite by accident...

In *Scarborough (Borough) v. R.E.F. Homes Ltd.* (1979), the Ontario Court of Appeal described a tree trespass, where a contractor cut trees on a municipal road allowance, as "completely inadvertent," but nonetheless awarded the appellant $4,000 plus the costs of appeal.

In *Richardson v. Avery* (1995), the Court found the trespass and the tree removal unintentional because the property line was poorly marked and the trees were growing in a natural area. The plaintiff was awarded $6,858.88 in damages.

The defendant in *Boudreau v. T.P.Downey & Sons Ltd.* (1987) mistakenly instructed his employees to cut one third of the trees on the property of the plaintiff. In this case, five to twelve cords of pulpwood and logs were removed from the property. The trees that the defendant was contracted to cut were located 107 metres away from the property line of the plaintiff. The Trial Court awarded the plaintiff $7,126. This award was reduced to $376 by the Appeal Court.

The defendant in *Sinkewicz v. Schmidt* (1994) unintentionally bulldozed the trees on a vacant property owned by the plaintiff, who was awarded $3,400 in damages.

3.4.2 Miscommunication

Occasionally, the owner of one or more trees may grant permission for some minor trimming to abate the nuisance of overhanging branches, only to find that the trimming work has removed the entire tree. Such was the case of *Forbes v. Griffin* (1989), where verbal permission was given for some trimming of willow trees, but the neighbour went over the property line and cut them down to three foot high stumps. The defendant deliberately trespassed and caused damage assessed by the Court at $1,500.00.

3.4.3 Property Improvement

One of the more common reasons defendants trespass is to improve the conditions of their own lot. Trees have been cut to increase exposure to sunlight, improve views, facilitate construction, remove a potential hazard, and to improve the growth of selected trees (by thinning the undesired species).

Sunlight Exposure

In *Kates* (1989), the defendant trespassed onto a neighbouring 1.2 hectare property while the owners were away and cut down 13 hemlocks, each measuring approximately 12 metres tall. The objective of the trespass was to provide the defendant's pool area with more sunlight. The Court awarded the plaintiff $21,000 for replacement costs, $13,000 for loss of amenity, and $26,000 in punitive damages.

In *O'Cadlaigh v. Madiuk* (1994), the defendant cut the neighbour's trees to permit more sunlight to enter his own property, thereby making it more attractive for resale. When the lot was subsequently sold, the advertisement emphasized its sunny character. The Court awarded the plaintiff $10,000 in general damages, $4,800 for clean up work, and $ 25,000 in punitive damages.

In *Goodwillie*, the defendant removed trees to expose a lawn area to more sunlight and to improve his view to the south and southeast. The plaintiff was awarded $9,027 in general damages.

View Improvement

There are many examples where trees have been cut, topped or limbed to create or improve views on private land. In *McKim v. Degraaf* (1991), the defendant was removing trees that he thought were on a highway right-of-way and could be cut with impunity. During this tree cutting, he deliberately went onto the plaintiff's property and cut down three red alder trees. The defendant admitted his transgression and the Court awarded $3,000 in general damages and $2,000 in exemplary damages because the trespass was willful.

In *Cordiez,* the defendant cut 28 trees to get a better view of the Kokanee Range of mountains in the West Kootenay region of British Columbia. The defendant said that generally where a tree broke his skyline he cut it down regardless of whether it was on his property or the plaintiff's. The plaintiffs were awarded $6,700 in damages.

The defendant in *Horseshoe Bay Retirement Society* (one of the more celebrated trespass cases in recent years) cut and limbed trees to achieve an ocean view of Horseshoe Bay and Howe Sound, B.C., knowing that such a view would increase the selling prices of the proposed houses by 10% or $68,000. The Court awarded the plaintiffs $33,172 in general damages and $100,000 in exemplary damages.

...the Court allowed the trimming of trees because it had occurred previously on the property

In one case, *Urquhart v. Sewell* (1991), the Court allowed the trimming of trees for view improvement because it had occurred previously on the property. In this case, the plaintiff sold a water view lot adjacent to the defendant's lot. Together the parties, with a tree trimmer, identified a number of trees which were to be topped or trimmed in a view corridor on the plaintiff's lot so the new purchasers should have the view that they had been sold. The view corridor was trimmed four times over a period of several years. The fifth tree trimming along the view corridor, which was the subject of the court action, was no more extensive than any of the previous occasions, and no trees were topped or trimmed other than those first agreed upon by the parties. The Court dismissed the action with costs to the defendants, and assessed no damages as the result of the trimming of the view corridor because the trimming had taken place before.

Tree removal to create an ocean view has occurred on lands owned by the public too. In *Prince Rupert (City) v. Pederson* (1992), the defendant, Mr. Pederson, cleared an area approximately twice the size of his lot on the adjacent city land, which was a forested park. A Prince Rupert city planner saw Mr. Pederson, cutting and removing the trees on a Sunday. Asked what he was doing, the defendant replied that he was not going to construct "a $200,000 home without a view." The defendant said that he was cutting the trees down on a Sunday since city employees would be off duty on that day and everything could then be cleared away. Mr. Pederson also stated that he would remove the trees by driving copper nails into thc wood or by dumping diesel fuel around their roots if necessary. The police attended the incident and the tree cutting was halted. On subsequent dates more trees were cleared, and debris piled on to city land. At one point, Pederson was found operating an excavator in the park in an attempt to grade the city property and his land. City officials endeavored, unsuccessfully, to halt Pederson's activities. He ignored verbal and written

Tree removal to create an ocean view has occurred on lands owned by the public too

warnings from civic officials and from the police to leave the trees on city property alone. All to create a view of the sea! The trial Court awarded the plaintiffs $73,365 in general damages and $58,500 in exemplary damages. These figures were reduced by the appeal Court to $63,365 general damages and $25,000 in exemplary damages.

Construction Facilitation

Building construction provides many examples where trespass has occurred. Thirty five trees were cut beyond the property lines to clear the land for building purposes in *Hik v. Kerr* (1992). This action cost the defendant $31,815.

In *698419 Ontario Ltd. v. Biggs* (1994), the defendant cut a 10.6 metre Scots pine tree growing on his neighbour's land so that he could build a shed closer to the property line; a mistake that lead to an award for the plaintiffs of $4,210.

In *Lee v. Hersh* (1993), the defendant advised the plaintiff that in order to build his new home adjacent to her heritage home according to the approved plans, he would have to temporarily encroach on her lands and remove a few trees. As a result of the defendant's work, substantial damage was done to the plaintiff's property. The 3.7 metre deep excavation for the defendant's underground parking garage encroached 3.6 metres into the Lee's land. The excavation undermined part of their asphalt driveway and a concrete sidewalk. The wall of the plaintiff's garage and the corner of her home were exposed and undermined as well. Six trees were cut down, including four mature elms. The plaintiff was awarded loss of income at $16,000, special damages of $70,000, and punitive damages of $23,000.

Sometimes trees have been removed because the defendant wished to subdivide the land. In *Dykhuisen v. Saanich* (1989), trees were cut down on the city owned boulevard in front of property the defendant planned to subdivide. Evidence presented to the Court showed that just before noon on the day in question, an employee of the Municipality of Saanich observed a number of trees lying on the road and the defendant directing traffic around these trees. The Municipal employee instructed the defendant that permission was needed to cut down the trees. The defendant said, "I give myself my own permission." The defendant indicated that the trees needed to be removed to enable a survey and to provide access to certain of the lots he intended to create on his own land. The defendant was ordered to pay $6,371 in general damages and $1,000 in exemplary damages. He appealed but the appeal was dismissed.

Potential Hazard Removal

In *Ursel v. Jenkins* (1988), trees were cut down because the defendant was concerned about the hazard they posed. The defendant thought the trees might endanger the house he proposed to build, so he adopted a self help approach to solve the problem. This approach cost him $7,500 in exemplary damages and $5,000 in general damages.

Growth Improvement

In a case from Quebec, *Howard v. Morgan Trust* (1983), trespass occurred when birch trees were cut on a wooded lot to improve the growth of other trees growing there. The plaintiffs were originally attracted to the property by its location, convenient for commuting to Montreal. They were also attracted by the stands of gray and white birch, white pine, old pine trees, and isolated specimens of other species, all of which provided a pleasant open forest. At the time the trees were cut, the plaintiffs had not yet built their house and were using the land as recreational property. On the day in question, the plaintiffs discovered that trees on their property had been cut by employees of the defendant, Timmerlinn Ltd., who had been engaged by the co-defendant, Morgan Trust, to cut birch trees on its property to improve the growth of the pine trees on their land, adjacent to the land owned by the plaintiff. The Court ordered the defendant to pay $24,132.94 in damages.

3.5 Establishment of a Defence in Trespass

In a court action for trespass where trees have been cut, damaged or removed, defence is often limited to the following three themes:

1. the trespass was an inevitable accident;
2. the trespasser had prior consent; and/or
3. there was some other justification for the trespass.

In *Barnstead v. Ramsey* (1996), the tree cutting was not intended, but occurred due to a lack of clear communication between the defendant and their contractors, thus leading, inevitably, to accidental trespass.

In the second type of defence, a consent to conduct work on another property has to have been clearly given. Trespassers often believe (or argue after the fact) that they had consent to remove the trees. *Cordiez v. Muffly* (1991) and *Ursel v. Jenkins* (1988) provide two such examples. In *Cordiez*, the Court, in discussing the liability of Mr. Muffly, who had cut the trees, said:

> While he may have done so in the mistaken belief that he had the right to do so, his actions cannot be justified and I must therefore find the defendant…liable for the trespass and the resulting damage.
>
> It is not enough that Mr. Muffly believed he had tacit approval. The plaintiffs had the right to use and enjoy their lands in their pristine quality. They have been denied that inherent right by Mr. Muffly's arbitrary and deliberate conduct.

In *Kates* (1989), the defendant asked the real estate agent responsible for selling the property and a neighbour who kept an eye on the property while the owners were away about tree removal. The defendant knew he did not have the owner's consent to cut the trees. The defendant thought that, since the Kates had decided to sell the property, they might not care or might not even find out if the trees were cut. The defendant further thought that, once the property was sold, the new owners wouldn't want the trees cut. Justice J. Taylor stated:

> It is not a case, on one hand, of damage maliciously done, nor is it a case, on the other

> hand, of trespass by oversight, or in mistaken belief as to the existence of permission. It is a case of deliberate invasion of a neighbour's property rights done in the hope that the owner was no longer interested in the property, and that the trespass would pass unnoticed or without objection. Mr. Hall decided to take his chances, knowing he was acting unlawfully.

In the third type of defence, a trespasser may be able to justify the trespass in order to preserve property or because of a life threatening emergency.

Sometimes the defence has pleaded that the time between the court action and the trespass and tree damage has been too great. The defendant in *Howard* pleaded protection by a statutory limitation Act, but was not granted it.

In *Nailen v. Powell River (Municipality)* (1978), the issue of trespass was clouded by a dispute over ownership of the land where the trespass occurred. The defendant in this case pleaded that there was no trespass because the storm sewer was constructed on land not owned by the plaintiffs. The defendant also attempted to rely on a subsection of the Municipal Act that stipulated that a court action must be started within one year of the trespass. However, in light of the Municipality's lack of timely response to the damage claim, the Court disallowed the time period limitation defence, and found the defendant had indeed trespassed.

3.6 Locating the Property Line

Many of the cases cited in this chapter involve disputes about the exact location of property boundaries. Trespass, whether inadvertent or deliberate, hinges on the location of the property line. Contractors are particularly vulnerable to miscommunication regarding the location of the property line, since they are less familiar with the land. The courts have repeatedly found that the responsibility for determining property line location lies with both the client and the contractor. In some cases the client has argued that the contractor did not listen and was therefore at fault. In others the contractor has simply followed the line identified by the client, assuming it to be correct. Contractors have also on occasion used their own experience to try to determine the location of property lines.

The courts have repeatedly found that the responsibility for determining property line location lies with both the client and the contractor

The courts have ruled that the contractor has a duty of care to locate the property line so as not to trespass onto an adjacent landowners' property during the course of work. In *Cosco v. Scott* (1984), the Court determined that the defendant, High Tree Services, was in the business of cutting and pruning trees. High Tree Services may have been misled as to the exact property line by the lay of the land and the assurances of the defendant Davis, but the circumstances called for an exact determination of the boundary. No such determination was made. For this reason the co-defendant, High Tree Services, was found liable to the plaintiffs for the wrongful cutting of their trees. Similar issues arise in *Sans v. Brassington* (1997).

In *Barnstead,* three fir trees on the plaintiff's property were cut 5 metres from the boundary line and other trees were damaged. Several large alder and native shrubs adjacent to the boundary had been removed. Large holes were made along the property line when the alder stumps were taken out. The Court held the tree clearing contractor as well as the defendant liable. The defendant had a responsibility to determine and mark the precise boundary line prior to tree removal on his property. The carelessness of the defendant caused substantial tree damage.

In *Howard*, the co-defendant Morgan Trust was absolved of blame. The tree clearing contractor was assessed all damages in the action for trespass. Morgan Trust hired the services of the co-defendant company to cut trees on its property. The co-defendant Timmerlinn Ltd. cut trees on the adjacent property, which belonged to the plaintiff. The Quebec Superior Court upheld the ruling of a lower court, which found that the co-defendant Morgan Trust was not liable because an employer-employee relationship did not exist between Morgan Trust and Timmerlinn. Timmerlinn was seen by the Court as in independent contractor in this action. Therefore, Morgan Trust was not at fault. The co-defendant Timmerlinn Ltd. admitted liability and received compensatory and punitive damages.

...the co-defendant was not liable because an employer-employee relationship did not exist

In *Mills v. MacAleese* (1987), the New Brunswick Court of Appeal found the tree clearing contractor liable to the plaintiff neighbour for trespass and awarded damages.

In addition, the Court found that the contractor was entitled to recover the damages assessed from the co-defendant because he made a negligent mis-statement in describing his property lines to the contractor. The owner of the property had a duty of care to both the clearing contractor and to his neighbour whose lands suffered the trespass.

In *Symons v. Madden* (1987), the boundary line was disputed after a tree was cut down. Both parties retained their own surveyors to determine the proper location, only to find that the surveyors could not agree. The final location was decided by a judge who reviewed the survey procedures to determine which was more accurate, and determined that there had been an accidental trespass in cutting down the tree.

The owner of the property had a duty of care to both the clearing contractor and to his neighbour whose lands suffered the trespass

Trespass also took place in *Verkerk v. McDowell* (1996). The logger measured the back property line, then sighted down the side line to a point believed to be another corner. He left extra wood to avoid a trespass, but in fact he did cross the line and remove some of Mrs. Verkerk's trees. The judge noted that:

> The responsibility for accurately determining the boundaries when logging or doing clearing work lies with the owner doing the work and his contractors. Work done off the property is a trespass even if done accidentally and in good faith.

A similar situation occurred in *Horton v. Speta* (1996), where again an attempt to estimate the approximate location of the property line proved to be incorrect, and trespass took place.

Criminal courts are no less tolerant of trespass than civil courts. In one of the rare criminal cases, *R. v. Laurie Richmond* (1995), a criminal court action was successfully pursued against Mr. Richmond, who lived next to Lighthouse Park (a municipal park in West Vancouver, British Columbia). Mr. Richmond had a local tree contractor cut down three large western red cedar trees in the park. Mr. Richmond's defence that he believed the trees to be on his property was not upheld at trial and he was found guilty of mischief to municipal property. He was fined $3,000.00 (with three months in jail if he did not pay the fine), plus an additional $24,500.00 in compensation. Interestingly, the contractor was not found liable in this case.

3.7 Summary

The reasons for trespass are varied but contain a common thread: the defendant either mistakenly or willfully trespassed onto the plaintiff's land and cut, damaged, or removed trees and shrubs. One of the most common reasons trees are damaged during trespass is to improve the conditions of the property owned by the defendant; for example, to get more sunlight, improve views, or facilitate construction. Many different types of awards have been made by the courts in cases involving the cutting, damage, or removal of trees in Canada.

Anyone involved in site clearing, tree pruning, or any other form of site preparation, whether as a contractor or as an owner or an owner's representative, has a very strong duty of care to accurately locate the boundary lines of the property. For the practising arborist, knowing the boundary line location is essential. It should be questioned carefully, and noted in writing before any work is undertaken. If there is any doubt, the location of the boundary line must be clarified before the work is undertaken if trespass is to be avoided.

Chapter 4
Trees Growing On Boundary Lines

4.1 Introduction

Trees growing on boundaries are of special interest due to the potential for disputes over who owns what and to what extent. Late 19th century and early 20th century court actions originating in Great Britain have played an important role in defining boundary or border trees. They have also provided several important precedents which the courts in Canada have followed.

Trees that are growing on boundary lines, or border trees, are those whose trunks, at ground level, grow either on or beside the property line, and which have roots or branches growing into adjoining properties. When the trunk grows partly on each property, the two property owners are tenants in common of the tree.

Trees that encroach on a boundary line may become a nuisance to an adjacent land owner. In such cases, in common law, the owner of the property is entitled to take measures to abate that nuisance.

4.2 Understanding Nuisance

The law of nuisance is highly relevant to boundary trees. In many cases trees have been damaged because they were close to a boundary, and the branches or roots beyond the boundary were considered to be a nuisance.

Nuisance takes many forms and can be quite complicated. It is broadly defined in Black (1990) as:

> Nuisance is that activity which arises from unreasonable, unwarranted or unlawful use by a person of his own property, working obstruction or injury to right of another, or to the public, and producing such material annoyance, inconvenience and discomfort that law will presume resulting damage.
>
> That which annoys and disturbs one in possession of his property, rendering its ordinary use or occupation physically uncomfortable to him.
>
> An offensive, annoying, unpleasant, or obnoxious thing or practice; a cause or source of annoyance, especially a continuing or repeated invasion or disturbance of another's right, or anything that works a hurt, inconvenience, or damage.

Nuisance can be private where the act or practice affects only private lands, or public where it affects many people or all residents in an area. Some instances cite both types of nuisance. Private nuisance typically involves someone undertaking, or permitting to be undertaken, an activity that within the boundaries of his property is quite lawful, but which, if it crosses over the boundary line, may:

1. become an encroachment resembling trespass on the neighbouring land;
2. cause physical damage to the neighbour' buildings, land or vegetation; or
3. may interfere with the neighbour's enjoyment of his property.

Possible remedies for nuisance are: outright damages based on an assessment of the damage resulting from the nuisance; or an injunction that the nuisance must be abated and not allowed to recur. To issue an injunction there must be actual or anticipated damage proven. In cases where damage has not yet occurred but is anticipated, it may be possible to obtain an injunction if there is clear evidence that the damage anticipated would be substantial and irreparable. Typical nuisance cases involve trimming back overhanging branches and underground roots.

One of the first recorded cases concerning border trees in English common law was *Crowhurst v. Amersham Burial Board* (1874). The defendants had planted a yew tree 1.2 metres from the property line. The yew grew over the boundary line and into the meadow of the plaintiff, whose horse ate the overhanging yew branches, was poisoned, and died. Since there was no precedent, the Court followed several previous cases where it had been ruled that a person who brings onto his land something which is likely to do damage is responsible if it escapes and damage occurs. The defendants were found liable for damages because the yew branches escaped over the boundary line.

To issue an injunction there must be actual or anticipated damage proven

The case of *Lemmon v. Webb* (1894) is also important in that it defined what may be done to overhanging branches and roots. Several large old trees, which had branches overhanging the defendant's property, were growing on the plaintiff's land along the boundary line. The tree branches had grown over the boundary for more than twenty years. The defendant, Webb, without notifying the plaintiff and without trespassing onto the plaintiff's land, cut off a number of branches to the boundary line. Initially the Court found that Webb was wrong to lop off the branches without notifying the tree's owner of his intent. Webb appealed, and the appeal judges reversed the decision, noting that:

> The owner of a tree has no right to prevent a person lawfully in possession of land into or over which its roots or branches have grown from cutting away so much of them as projects into or over his land, and the owner of the tree is not entitled to notice unless his land is entered in order to effect such cutting. However old the roots or branches may be, they may be cut without notice, subject to the same condition. The right of an owner or occupier of land to free it from such obstructions is not restricted by the necessity of giving notice so long as he confines himself and his operations to his own land, including the space vertically above and below its surface.

...the principles of Lemon v. Webb have been in effect as common law since 1894

An appeal of the appeal court's decision was dismissed and the principles of Lemon v. Webb have been in effect as common law since that time.

Damage to the property of the plaintiff occurred in *Smith v. Giddy* (1904) and set another important precedent. Elm and ash trees were cut because they overhung the property of the defendant and interfered with the growth of his fruit trees. The judge noted that *Crowhurst* was the only court action for damage in nuisance caused by overhanging trees, but the damage in that case was of a different kind. At the time of *Smith,* there was no precedent in English common law regarding overhanging limbs of non-poisonous trees which caused damage. The Court ruled that when trees were overhanging the boundary and not doing any damage, the defendant's only right was to cut back the overhanging portions, but where the limbs were actually doing damage, the defendant had a right to sue for damages under the law of nuisance.

...where overhanging limbs are actually doing damage, the defendant has a right to sue for damages under the law of nuisance

In an interesting twist, *Mills v. Brooker* (1919) dealt with whether fruit growing on overhanging branches could be picked and sold by the neighbour. The Court found that the fruit belonged to the owner of the tree, regardless of whether it was still attached to the tree. Moreover, although the defendant had a right to trim the overhanging branches (following *Lemmon*), he could not rightly pick the fruit for his own use. Similarly, branches severed to abate a nuisance could not be converted into cash for the benefit of the neighbour whose property they overhung. Rather, they had to be returned intact to the owner. The case was appealed, but dismissed.

As a result of these early precedents, the law of nuisance with respect to overhanging branches and roots is reasonably clear. Overhanging branches or roots which are not damaging are considered a nuisance only. Therefore, the property owner has the right to prune them back to the property line without giving notice to the neighbour who owns the trees. Common prudence, though, suggests it would be wise to at least notify the neighbour of the intention to prune.

However, if the roots or branches are damaging an adjacent property, and the owner of the tree is unwilling to mitigate or compensate for the damage, a court action may be pursued. The plaintiff must prove the damage exists. The owner of the trees has no right to put an adjoining land owner to the expense of abating the nuisance. The age of the trees and the length of time they have been overhanging or encroaching the boundary are not relevant. Property owners are entitled to the full use and enjoyment of their land free from nuisance emanating from neighbours.

> ***The tree owner has no right to put an adjoining land owner to the expense of abating the nuisance***

Three key issues emerge. First, roots or branches may be pruned only to the property line. If pruning crosses the property line, no matter how slightly, it becomes trespass. Secondly, it is not clear what happens if the pruning work subsequently leads to the death of the tree. It was noted in *Anderson v. Skender* (1993) that "…it is now accepted throughout the common law world to which we belong that an obligation rests on everyone to take reasonable care not to cause needless injury to the property of others." Thirdly, fruit which has fallen from overhanging branches cannot be retrieved by its owner without permission from the neighbouring property owner on whose land the fruit lies.

4.3 Damage to Trees on Boundary Lines

Trees growing on boundary lines have been cut down, had branches removed at or beyond the property line, and/or had their roots severed. The precedents noted in the previous section have defined what is acceptable in Canadian law and are therefore widely cited.

4.3.1 Tree Roots Extending Beyond Boundaries

When tree roots extend under a neighbour's house, they can do considerable damage by sucking water out of the soil, causing it to shrink and possibly also causing the house's foundations to subside. This problem was addressed in three cases: *Butler v. Standard Telephones & Cables Ltd.* (1940), *McCombe v. Read* (1955), and *Davey v. Harrow Corporation* (1958). In *Butler* the roots of Lombardy poplar trees growing under the plaintiff's house removed water from the clay soil, which shrank, causing the wall of the house to crack and show signs of settling. The plaintiff, Butler, recovered the cost of the damage.

In *McCombe,* the roots of Lombardy poplar trees encroached onto the plaintiff's property, and undermined the foundations of his house. As in *Butler,* the trees had withdrawn sufficient moisture from under the house foundations to cause the clay soil to shrink and the foundation to settle. A trench was dug around the south and west sides of the house to a depth not less than 91 centimetres and to a maximum of 2.74 metres. Fibrous roots and roots up to 2.5 centimetres in diameter were found growing adjacent to or under the house foundation. Some roots had grown entirely under the house. The Court ruled that damages were payable but not adequate. The roots would continue to grow and remove water from the soil causing additional damage in the future. The plaintiff would have to return to court again and again. This eventuality was viewed as unfair to the plaintiff. The plaintiff was awarded an injunction to stop the defendants from allowing the roots of their trees to encroach onto his land.

...when tree branches or roots encroach and cause damage...the defendant is liable for damages

In *Davey,* the plaintiff's house was damaged by encroaching elm roots which caused his house to subside. Once again, the damage was caused by tree roots absorbing water from beneath the foundations of the plaintiff's home and garage. The Court held that when tree branches or roots encroach and cause damage, as in this action for nuisance, the defendant is liable for damages.

Encroaching tree roots do cause other kinds of problems, too. For example, the roots of 14 Lombardy poplar trees encroached across the boundary line in *Mendez v. Palazzi* (1976), producing suckers that destroyed the plaintiff's lawn and disrupted the patio. The plaintiff estimated that she spent 100 hours per year removing the profuse suckers and roots from her garden and patio. She was able to establish an actionable nuisance and therefore was entitled to damages.

Crosty v. Burlington (City) (1978) is an example where liability in negligence was not awarded because the homeowner had assumed a duty of care. A city tree had on numerous occasions blocked a lateral sewer pipe connecting the city's main pipe to the plaintiff's home. Until city policy changed in 1962, the plaintiff had got the city to come and unplug the pipe. Following the policy change, the plaintiff was loaned some city equipment to use in clearing the blockage himself. Nevertheless, he subsequently had to purchase his own equipment. In 1968 the city installed a new lateral pipe. Some months later the new pipe was once again blocked with roots, and the plaintiff used his own equipment to unplug it. In July of 1974 the pipe became blocked once more, backing up and flooding the basement of the plaintiff's house with sewage.

...liability in negligence was not awarded because the homeowner had assumed a duty of care

At issue in this case was whether the city had a duty of care to maintain every single lateral sewer pipe within its jurisdiction or whether its policy of relying on homeowners to notify them of a blockage was sufficient. The judge found that, although the *Local Improvement Act* (Ontario) imposed a "general duty rather than a compulsory standard of care," it did not impose an "absolute liability for failure to repair or keep in repair [items such as sewer pipes]." The judge noted that the city was not negligent in complying with its statutory obligations because its policy was to rely on citizens to report problems, because the plaintiff had helped himself over a number of years, and finally because the plaintiff had latterly failed to notify the city of the problem. Referring to the case of *Jordan House v. Menow and Honsberger* (1974), the judge quoted Laskin, J. as follows:

> Moreover, in considering whether the risk of injury to which a person may be exposed is one that he should not reasonably have to run, it is relevant to relate the probability and the gravity of injury to the burden that would be imposed upon the prospective defendant in taking avoiding measures.

In other words, if the probability and potential gravity of injury is slight, is it reasonable to expect a standard of care that encompasses all possible levels of risk?

Buie v. Etobicoke (City) (1984) is an example where the defendant, the City of Etobicoke was found negligent for not acting soon enough to address a complaint. The city had a policy for dealing with complaints about roots blocking sewer pipes, but in this case delayed addressing the problem for four months. According to Justice Caswell, the city had a duty to

> take reasonable steps, once it was notified of the problem and before the growing season commenced, to girdle the tree....Once the City knew that the tree was...causing physical damage, there was a duty on it to take reasonable steps to prevent further damage....The City failed to take such steps, although it had the time and resources to do it.

In the case of *Canadian Red Cross Society v. Past* (1979), part of the claim alleged that tree roots had damaged a nearby building. In the absence of substantiated evidence, that part of the claim was not allowed. In *Anderson* (1991), when the defendant (a contractor) built his house, his equipment operator excavated to the property line for the building foundation and 50% of the root system of a border tree was destroyed. The defendant was found liable for his actions; the outcome of this case is explained in detail in Section 4.3.3.

4.3.2 Tree Branches Overhanging Boundaries

Examples where overhanging branches have been trimmed to and beyond the property line, resulting in a claim of trespass include *Stowe v. McTaggart* (1993) and *Winger v. Torkelson* (1989), both of which are discussed in the next section. In the case of *Canadian Red Cross Society*, damage was claimed as a result of on overhanging elm branch which damaged an adjacent building.

4.3.3 Trees Straddling a Boundary Line

The courts have treated trees which actually stand on the boundary line differently from those which extend only roots or branches over the boundary. Several cases demonstrate how an attempt to remedy perceived nuisance or potential damage can create large problems.

In *Loverock v. Webb* (1921), a large alder tree leaned over the airspace of the defendant. The trunk was entirely on the property of the plaintiff. The defendant cut the tree trunk some distance from the ground, where he thought the property line might be located if a plumb bob were hung from the trunk. He erred and cut the trunk a few centimetres over the property line. The tree was destroyed. Nominal damages were awarded since the trespass was viewed by the Court as a technical trespass. Nominal damages were also assessed in *Winger v. Torkelson* (1989) for a technical trespass.

The case of *Anderson v. Skender* (1991) provides comprehensive guidance on the issue of boundary trees and damage. The Skenders bought a vacant lot next to the Andersons. Mr. Skender planned to build a house on his lot, and wished to cut three Douglas-firs standing either entirely or mainly on the Anderson side of the boundary. Two of the trees had double trunks, one trunk of each leaning into the Skenders' airspace. At ground level, one of these trees was entirely on the Anderson property. The other one, while mostly on the Anderson property, extended eight centimetres over the line into the Skenders' property. The Andersons and the Skenders were tenants in common of this tree. The third tree had one stem which stood entirely on the Anderson side of the line. Close to half of the root systems of each of the two double trunk trees, and about 20% of the third tree, extended into the Skenders' property.

Anderson v. Skender (1991) provides guidance on the issue of boundary trees and damage

The defendant, Skender, was of the opinion that the trees obstructed the construction of his new home. The Andersons suggested that a Board of Variance ruling might allow the Skenders to re-site their proposed house so that damage to the trees would be minimized during construction. The Andersons wrote a letter to resolve the problem but the Skenders, in the words of the trial judge, "answered the plaintiff's letter with a chainsaw." The defendant trespassed and cut two of the trees, leaving one trunk of each of the twin stemmed trees standing on the Andersons' property. These two remaining trunks were seriously weakened and, upon the advice of an arborist, the Andersons removed them because their survival was unlikely. The third tree suffered only root damage, which occurred during the excavation work.

> ***...the two property owners were tenants in common of the border tree***

Mr. Justice J. McKenzie ruled that the two property owners were tenants in common of the border tree, citing in his judgement the early American case of *Griffin v. Bixby* (1841) where it was ruled that

> a tree standing directly upon the line between adjoining owners, so that the line passes through it, is the common property of both parties, whether marked or not, and that trespass will lie if one cuts and destroys it without the consent of the other.

With regard to the severing of main trunks and substantial root injury, Justice McKenzie stated:

> Cutting overhanging boughs is quite a different thing from severing one of two trunks and damaging 50 percent of the root system. One has to look at it, as it were, from the tree's point of view. It is an assault on the general health of the whole tree, affecting as it does the energy creating foliage and the energy storing roots. If the assault is carried far enough it can be fatal to the tree.

The defendants had an interest as tenants-in-common of the main double tree. Their interest in the other two trees is tenuous but nevertheless they had some interest because of the root penetration under their land. The incontrovertible fact, with respect to the two double trees, is that the defendants had no legal right to effectively destroy them and no absolute right to damage the third tree.

The Skenders contended that their right to self-help included a right to destroy the invading roots and trunks of the plaintiff's trees, even if it caused the trees to die. In this case, the defendant argued that the right to self-help, and the way it was exercised, was not limited by, or subject to, any sort of duty of care.

In response, the Court of Appeal (*Anderson v. Skender,* [1993]) said:

> [T]he view that a landholder has an absolute right to destroy a neighbour's border trees, by the needless infliction of mortal injury on those portions which extend into the former's property… I will for now say no more than that this view of the law ought not … to receive gratuitous support or encouragement.

Mr. Skender was assessed $6,136 for the value of the trees, $2,000 in punitive damages, $250 for damage to the driveway, and $289 as the cost to remove the remaining tree. The Court of Appeal reduced the award by $436.18, disagreeing with the lower court's award for the root damage to the third tree because Mr. Skender did not trespass onto the Andersons' property to inflict the injury suffered by this tree.

In a court action from Nova Scotia, *Peters v. Dodge* (1910), the Court of Appeal ruled that, in common law, any tree growing on the boundary line between adjacent properties is common property, and consequently can not be cut down without the consent of both owners. In this case, a border oak tree 51 centimetres in diameter was cut down by the defendant.

In *Stowe v. McTaggart* (1993), the roots and trunks of a 4.9 metre high laurel hedge straddled the property line. In this case, the defendant reduced the height of the hedge by 2.7 metres to improve his view. Mr. McTaggart trespassed onto the plaintiff's property and removed sufficient growth to jeopardize the survival of the hedge. The plaintiff was forced to remove the hedge and replace it with a privacy fence and cedar hedge.

...any tree growing on the boundary line...can not be cut down without the consent of both owners

The plaintiff was awarded $3,263.63 to build the fence and plant a hedge, plus $2,000 for the loss of amenities, loss of privacy, and for the disturbance caused by the defendant. The defendant counterclaimed that the plaintiff's workers caused considerable damage to his property when they dug out the remnants of the laurel hedge and built the privacy fence. The Court agreed that, in law, a trespass had occurred, but it was the result of the original actions of the defendant. The defendant was awarded the nominal sum of $1 for damages arising from the counter claim.

In *Michaels v. Zwickel* (1994), the defendants' construction contractor removed three acacia trees and some bamboo and laurel shrubs growing on the boundary line in order to construct a fence and build a deck. The construction workers trespassed and built the new boundary fence 65 centimetres inside the plaintiff's property. The trial, originally scheduled for not more than two days, ballooned into a major cause célèbre. It required 10 days of the Court's time and was heard over a one and half year period. The credibility of the plaintiff, defendants, neighbours, and others was challenged during the court proceedings. The Supreme Court of British Columbia awarded $4,000 for the loss of amenities, $4,863.80 for the cost of replacing the trees and shrubs removed, and $4,000 in punitive damages. In this case, the judge suggested that the defendant's conduct was outrageous and high-handed from the time the construction workers trespassed onto the plaintiff's property until the defendant was finally forced to remove the fence.

In *Chan v. Rowley* (1992), the defendant, Rowley, cut down three mature cedars straddling the boundary line which he then milled into lumber to construct the deck of his summer home. The Supreme Court of British Columbia awarded the plaintiff $3,000 for the loss of amenities and $4,000 for the cost of replacing the three cedar trees.

The defendant, in *Winger v. Torkelson* (1989), complained that the two large silver maple trees growing along the boundary line were irritating her because of the leaf litter which fell onto her property. Forty percent of the trees overhung the defendant's property.

While the plaintiff was away 5.4 to 6.0 metres of large limbs were removed from one main trunk without her consent. To add insult to injury, the tree cutter came to the plaintiff's house the evening of the removal and attempted to sell her the wood that had been cut. In this case, the Court ruled that the tree cutting was a technical trespass, because the property of the plaintiff was not damaged. The plaintiff was awarded the value of the limbs removed by the defendant, $315, which was the value of the wood as firewood. Similar issues arose in *Black v. Zager* (1982).

Because of differences in provincial legislation, not all cases use the same interpretation of the law

Because of differences in provincial legislation, not all cases use the same interpretation of the law. For example, in Ontario, the *Trees Act* (R.S.O. 1990) offers detailed guidelines about many different aspects of trees, and in part states:

> 2. An owner of land may, with the consent of the owner of adjoining land, plant trees on the boundary between such lands, and every tree so planted shall be the common property of the owners.
>
> 3. Every person who ties or fastens any animal to or injures or destroys any tree growing for the purposes of shade or ornament upon a boundary line between lands, or who suffers or permits any animal in his charge to injure or destroy or who trims, cuts down or removes any such tree without the consent of the owners thereof, is guilty of an offense and on conviction is liable to a fine of not more than $1000.

In *Graham v. Da Silva* (1984), Mr. Da Silva cut the branches of a maple tree back to the property line because of concerns that they would damage his garage. All of the tree trunk, with the exception of 1.3 centimetres, grew on the plaintiff's land. The tree leaned into the defendant's airspace, and after the limbs were removed, only a trunk 1.5 metres high remained. The defendant cut the trunk to the ground as well, while the plaintiff was away at work. The plaintiff claimed that Da Silva trespassed. The defendant counterclaimed in nuisance. Both the action and the counterclaim were dismissed by the Ontario County Court. The Court ruled that the maple tree was not a boundary tree as defined by the *Ontario Trees Act.* The defendant, therefore, did not require the plaintiff's permission to lop off branches on his side of the property line. The removal of the large trunk was, the Court considered, a technical trespass with no damage done to the lands of the plaintiff. Had the claim been allowed, the Court would have assessed damages at $1,000 as stipulated in the *Act.*

The defendant, therefore, did not require the plaintiff's permission to lop off branches on his side of the property line

In another court action from Ontario, *Centrum Land Corp. v. Institute of Chartered Accountants of Ontario* (1988), the defendant, the Institute of Chartered Accountants, wanted to construct an addition to its existing building. In order to construct the addition, a number of trees which were growing on the property line were removed. There was no evidence that the trees growing on the boundary line had been planted with the common consent of the owners of both properties. The Court ruled that in the absence of evidence, subsections 2 and 3 of the *Trees Act* did not apply. The Court said that the Institute was entitled to cut any branches or roots which interfered with enjoyment of its property, including the right to build on the property within 2.5 centimetres of the boundary line. The Court further said that, even if the trees were so close to the property line that the trimming of branches or roots limited their survival, there was no authority to suggest that the defendant must then suffer the nuisance. The Court in this case applied the ruling made in *Graham*. The trees were cut with no penalty to the defendants.

4.4 Summary

Late 19th century and early 20th century court actions originating in Great Britain have provided several important precedents which the courts in Canada have followed.

The law of nuisance in regard to trees growing on boundary lines is reasonably clear. A property owner is entitled to cut back to the boundary branches or roots of a neighbour's trees which extend over the boundary, although the costs of such cutting can not be claimed back from the owner of the tree. The tree trimmer is not legally required to solicit the consent of or provide notice to the tree owner. Nevertheless, requesting consent or notifying the tree owner of intention to trim is considered good practice. A property owner does have a duty of care to the owner of the tree being trimmed, and must ensure that the trimming does not create structural instability, other hazardous conditions, or the death of the tree in question. A land owner must also take reasonable care not to unnecessarily injure the property of the adjacent landowner when trimming is undertaken. All trimming work must not cross the property line, otherwise it will be considered a trespass.

When tree branches or roots extend into and damage an adjoining property, the property owner may seek damages in the courts as their right of action in the law of nuisance. When a tree trunk grows partly on each property, then the two property owners are tenants in common of the tree. If the property line runs through the centre of the tree, then each property owner has a 50% interest in the tree. When property owners are tenants in common, the tree cannot be felled by one property owner without the consent of the other, for it cannot be cut at ground level without the cut being at least in part on the other's lot. However, some provinces, such as Ontario, have legislation in effect that further determines what can be undertaken with boundary trees. Trees straddling property lines do not trespass; tree trimmers trespass when their work goes beyond the property line onto the land or into the airspace of the adjoining property owner.

Chapter 5
Trees Along Roads And Highways

5.1 Introduction

Trees growing along roads, streets, and highways have caused a number of different problems. Trees or branches have obstructed stop signs at street intersections, reducing their visibility and thereby causing serious traffic accidents. Similarly, injury and death have been caused by trees or parts of trees falling onto roads and highways. Tree roots have caused more mundane problems, such as property damage when they grow into drains under a road right of way or injury to people who trip on uneven sidewalks raised by roots. Trees growing along roads and highways have been cut down, bringing about court actions for damages.

5.2 Vegetation Obscuring Stop Signs

In *Ficht v. City of St. Catharines* (1978), the driver failed to see a stop sign because it was obscured by the branches of an elm tree. The defendant argued that, under certain provisions of the Municipal Act (Ontario), he could not be held liable and the case should be barred. Citing numerous cases, the judge ruled that in fact the action could proceed, since the Municipal Act did not apply in this case. Although the judge did not rule on who was liable, he did find that the placement of the stop sign in relation to the tree was wrong, rather than the situation or arrangement of the tree.

In *Greatrex v. Ennismore (Township)* (1984), Russell Greatrex was driving in a westerly direction along a road in the Township of Ennismore in Ontario. Upon entering an intersection, he collided with a car carrying the Jardine family. All four members of the Jardine family died in the collision. Several court actions arose as the result of the incident. Russell Greatrex alleged that the stop sign, directed toward westbound traffic on the township road, had been obscured by the branches and leaves of a tree. The issue in the court action was whether or not the two municipalities, the Township of Ennismore and the County of Peterborough, were negligent in removing the foliage, and if so, whether the negligence caused or contributed to the cause of the fatal crash.

Township road #9, upon which Greatrex drove, was oiled and gravelled. County road # 16 was a through highway and the intersection where the accident occurred was regulated by stop signs on township road # 9 only. The stop signs were designed to stop east and west bound cars prior to entering county road # 16. Russell Greatrex alleged that he saw the intersection, the stop sign, and the Jardines' car simultaneously. The judge concluded that a motorist was able to see the stop sign at least 45 metres from the intersection whereas the regulated distance was 60 metres.

In this case, the County of Peterborough erected the stop sign and had a duty to maintain it. The county considered it the township's duty to make sure the sign was visible without obstruction for the required 60 metre distance.

The Township of Ennismore argued that it should not be held liable if the sign was not visible. The township maintained that it had regularly inspected its roads and had not noticed the obstruction of the stop sign in question. Township workers, who were responsible for checking the roads, should have seen the obstruction and trimmed the branches. The judge concluded that the poor visibility of the stop sign and, therefore, the negligent non-repair of the township road contributed to the cause of the tragedy. The judge ruled that the township, not the county, was liable for the damages caused by its negligence. It was not reasonable to expect the county to trim foliage which was clearly on a township road. Russell Greatrex was also found negligent, but to a greater degree than the Township. He was assessed 75% at fault and the township 25% at fault. Greatrex was driving too fast on a road not known to him. He made no attempt to slow down at the intersection. He did not brake or take any other action to avoid the impending crash.

The Court found that there was a duty to maintain stop signs at street intersections...

In another example, *Campbell v. Calgary (City)* (1984), the Court found the city liable for negligence when a traffic accident occurred because a stop sign was obscured by vegetation. The Court found that there was a duty to maintain stop signs at street intersections, but that it had not been maintained with a regular routine of inspection.

The plaintiff in *Thornhill v. Martineau* (1987) was injured when her motorcycle collided with the defendant's van. The plaintiff ran through a stop sign at an intersection, claiming that she could not see the sign because it was hidden from view by the overhanging branches of nearby trees. The issue in this case was whether any duty of care was owed by the municipality to the plaintiff to make sure that the stop sign was visible at a reasonable distance from the intersection. The judge noted that there was no statutory duty imposed on the municipality to install or maintain traffic control devices, but that a municipal by-law authorized the municipal engineer to install and maintain traffic control devices "within the amount of funds provided by the Council for such purpose." The Court ruled that the way in which a municipal government chooses to exercise discretionary power such as the erection of stop signs is a policy matter which is beyond the review of the courts and which does not carry with it any duty of care which is recognized by the common law of negligence. However, the municipality could still be held liable if its stop sign maintenance fell under the category of operational activity. For reasons of efficiency and economy, the municipality had decided to maintain stop signs within its jurisdiction by trimming branches only in response to public complaint. This policy decision to not individually inspect each stop sign protected the municipality from liability because no duty of care arises from policy-making. Therefore, the Court dismissed the action.

The policy decision to not individually inspect each stop sign protected the municipality from liability...

In *Davidson v. Kamloops (City)* (1992), a similar court action took place. In this case, the plaintiff argued that the City of Kamloops was in breach of its statutory duty under the *Municipal Act* to repair and maintain the streets. The plaintiff went through a stop sign and collided with another vehicle which had the right of way. She claimed that she could not see the sign because it was obscured by tree branches.

The City of Kamloops had a computerized inventory system of all the stop signs. The signs were inspected twice a year to check their condition and visibility. In addition to these biannual inspections, the city also responded to complaints made by residents. In this case the city had made a policy decision to inspect the stop signs individually twice a year. The Court ruled that there was no evidence that the inspection of the stop sign at the intersection in question was negligent. The Court found the inspection procedure reasonable and as a result the city was found not liable. The decision was appealed in *Davidson v. Kamloops (City)* (1995), but the appeal was dismissed. The accident was not caused by the negligence of the City of Kamloops, but by the plaintiff's lack of attention to her surroundings.

5.3 Trees Obstructing Roads and Highways

Trees or parts of trees falling onto highways have caused many lawsuits over time. An early example from the United Kingdom is *Noble v. Harrison* (1926). A large branch from an apparently healthy beech tree fell onto a passing motor coach. The coach owner sued the estate owning the tree and won in County Court. The estate appealed and the appeal was upheld because the branch was only found to be defective once it had failed, and a competent prior inspection would not have been able to determine the presence of the defect in question. The fact that the tree had probably been planted (as opposed to being naturally sown) was of no relevance. A notation of subsequent importance in other cases, made by Justice Rowlatt, was that

> I see no ground for holding that the owner is to become an insurer of nature, or that default is to be imputed to him until it appears, or would appear upon proper inspection, that nature can no longer be relied upon.

"I see no ground for holding that the owner is to become an insurer of nature..."

This principle is revisited in several cases involving damage caused by falling trees, discussed later on in Chapter 8. *Noble* was affirmed in *Cunliffe v. Bankes* (1945) which involved very similar circumstances.

Several early decisions deal with municipal liability for road maintenance. For example, *Gilchrist v. The Corporation of the Township of Carden* (1876) involved municipal employees who cut a tree but left it lodged up in another tree. It subsequently fell down and killed a women passing along the road. After several iterations of court examination, the final ruling was that the municipality was liable: its employees had cut the tree, left it lodged in such a way that there was a possibility, no matter how remote, that it could fall onto the highway, and had therefore failed to exercise a reasonable standard of care in protecting users of the highway. A similar issue arose in *Steves v. South Vancouver (District)* (1897). Employees conducting other work on the road caused a tree to become unstable (in this case due to excavation of road fill) and again a fatality ensued. The district was found liable to the plaintiff for $10,000. The district appealed but lost on the basis that the employees had caused the problem and the Corporation, being responsible for the actions of its employees, must therefore be held accountable.

...the corporation, being responsible for the actions of its employees, must therefore be held accountable...

A tree falling on the road and causing personal and property damage was also seen in *Huestis v. Toronto (City)* (1926). In this case it was shown that the tree was plainly defective prior to the incident. Moreover, there was a by-law in effect stating "Trees …dead or partly dead …shall be removed as may be by the Parks Commissioner deemed advisable." The judge found that the city had a liability at Common Law, and by dint of its by-law had voluntarily assumed a duty of caring for the trees.

In *Ratkevicius v. R.* (1967), a high wind caused a large limb of a tree growing adjacent to a highway to fail. The branch crashed down on a car travelling along the highway, and injured the people inside the car. The trial judge ruled that the damage was caused by a sudden condition which the highway department could not have been reasonably expected to foresee—despite the fact that the tree had been noted as potentially dangerous by a patrolman, and subsequently by an arborist. Three months prior to the limb's failure, the arborist had recommended removing it. The Trial Court found the defendant, the Department of Highways, not liable for negligence. The case was appealed.

The Appeal Court overturned the lower court ruling on the basis that the Department of Highways was in breach of its statutory duty to maintain the highway in good repair, and was liable for the personal injuries that resulted. It had access to sufficient knowledge of the condition of the tree by dint of the staff arborist's report which was submitted three months prior to the incident. During those three months no steps were taken to abate the known hazardous condition of the tree. The appeal Court also increased the amount of the awards against the defendant.

> ***...the Department of Highways was in breach of its statutory duty to maintain the highway in good repair...***

In the case of *Wahl v. Listowell (Town)* (1988), municipal employees were taking down a tree. It fell on a nearby house and the owner sued for the damage sustained. The municipality attempted to have the case barred on the grounds that the Municipal Act provided, among other things, that no action could be brought for damages by the disposition of any tree not within the traveled portion of the highway. The municipality's attempt to have the case barred was dismissed on the grounds that the Act cited the disposition of trees, but in this case it was the falling of the tree, not its disposition, that was at issue.

An important question in such cases is the degree to which municipalities or other employers are responsible for the actions of their staff or appointed contractors. In the event of an accident, vicarious liability may be assumed by the employer and/or the contractor.

One of the more important actions regarding highways is *Just v. British Columbia* (1989). Mr. Just and his daughter were traveling Highway 99 en route to Whistler, B.C. for a day of skiing. A heavy snowfall stopped the traffic on the highway. While the Justs were waiting in their car for the traffic to resume moving, a rock weighing more than a ton careened down the rock slope and crushed their car. Mr. Just was injured and his daughter died.

The rock slope instability was due to a phenomenon called root prying. The roots of a tree had loosened the boulder from a natural slope above an engineered rock cut some 40 metres above the road. The highway department had a scaling crew whose job it was to visually inspect the rock cut faces. These inspections were made from the highway itself, unless there was a history of instability in the area or some evidence of slope failure such as fresh accumulations of rocks in the ditch alongside the highway. When there was fresh evidence, the cut was climbed and inspected.

In this case both the trial judge and the Court of Appeal dismissed the action on the grounds that the highway department system of inspection and its implementation were matters of policy. As a result, no liability could arise.

The case went to the Supreme Court of Canada where the appeal was allowed. The Supreme Court ruled that a new trial was needed to determine if the manner and frequency of the inspections were in fact not policy decisions but operational in nature. The Court found that the Crown owed a duty of care to users of the highway system to maintain these highways in reasonable condition. The new trial was ordered to determine what an appropriate standard of care might be, if it was breached in this instance, and if so, what the damages should be.

The Supreme Court ruled that a new trial was needed to determine if the manner and frequency of the inspections were...not policy decisions but operational in nature...

The new trial allowed the Court to determine whether the highway department had exercised its duty of care. Supreme Court Justice J. Donald stated:

> At the crux of this case lies the question whether there were sufficient indications of instability of the cut evident prior to the accident such that a climbing inspection was required. I am satisfied that if an engineer or member of the scaling crew had climbed and inspected both the constructed and at least the lower portion of the natural area, the danger presented by the tree would have been seen and the tree removed shortly thereafter.

The Judge found that indeed the highway department had breached its duty of care, and despite evidence that the rock slope was unstable, had failed to properly inspect it. Had the highway department done so, it would have noticed the tree and the loose rocks and could have taken steps to remedy the problem before an accident occurred. As a result Mr. Just was awarded damages to compensate for his severe injuries, lost wages, lost earning power, and lost pension benefits. In total, the damages amounted to $999,031, plus a management fee to help him administer the award.

In the Supreme Court of Canada case *Swinamer v. Nova Scotia (Attorney-General)* (1994), the plaintiff, Mr. Patrick Swinamer, was driving along Falmouth Back Road, a public highway in Nova Scotia. An elm tree, 18 to 21 metres tall and 76 centimetres in diameter crushed Mr. Swinamer when it collapsed across the cab of his truck. Mr. Swinamer became a paraplegic as a result of the accident. The elm was growing on farmlands 90 centimetres from the highway right of way. The tree was one in a line of elms whose branches extended over the center line of the highway. The tree had appeared healthy in the summer prior to the accident, but was badly infected with a fungus known as artist's conk. The tree failed at its base when it collapsed onto the road. The center of the tree at the ground level was pulpy, yet the wood outside appeared sound and decay was not evident there. A casual examination would not have pinpointed the infection, but an expert may have suspected the tree was disease ridden if an entry wound or a fruiting body of the fungus had been found at the base of the tree. To confirm a diagnosis of artist conk fungus, an expert would have had to bore the elm and evaluate a core sample from the tree.

The highway department regularly removed fallen trees and branches as part of their routine maintenance activities. In addition, trees that had been identified as hazards by the public or highway employees were removed. The divisional engineer in the county where the accident occurred had received complaints about dead trees standing along the highway right of way. There was no funding to take them down at that time, but the engineer authorized a survey of the right-of-ways to determine the number and location of these dead trees. The survey team marked 234 dead trees with survey stakes and reported their locations. Among the trees marked was a dead tree 122 metres from the one which failed and injured Mr. Swinamer. The elm which collapsed was not marked during the survey. Prior to the collapse of the elm, the engineer submitted a report, requesting money to remove the dead trees. Funding was granted for the removal of 66 of the 234 surveyed, with funding for additional removals made available in subsequent years.

The Nova Scotia trial judge found the highway department liable for Mr. Swinamer's injuries based on the judge's view that the department was negligent in the way in which it inspected the trees along the right-of-ways. The judge suggested that the department should have hired an expert who might have recognized a tree infected with artist's conk fungus and recommended it be taken down. The Nova Scotia Court of Appeal, when the action came before it, disagreed and dismissed the claim. The Court gave six reasons for its findings:

1. there is no statutory duty requiring the Minister to maintain provincial highways;
2. there is no liability imposed on an abutting owner for a nuisance on an adjoining property;
3. there is no statutory power to enter onto lands abutting the highway to inspect or remove trees;
4. any duty at common law to repair highways does not extend to adjoining lands;
5. even assuming a duty to remove dangerous trees from lands abutting a highway, there was no evidence that the tree in question constituted a danger prior to the accident; and
6. the finding by the trial judge that there was policy to inspect and remove diseased trees from adjoining lands, which was the key to his decision, was not supported by the evidence.

When the Court of Appeal judgement was appealed in the Supreme Court of Canada *(Swinamer v. Nova Scotia [Attorney-General]* [1994]), Justice J. Corey, speaking for the majority, disagreed with the first four conclusions but agreed with its overall conclusion that the action be dismissed. The highway department had made a policy decision regarding the inspection and removal of dead trees along highway right-of-ways. As a policy decision, it was immune from judicial review. The Supreme Court next determined whether the highway department was negligent in implementing its policy, and whether the operational work was defective. The Court found no negligence and ruled that it was not necessary for the department to hire an expert or to specially train the field surveyors to evaluate the trees growing along road right-of-ways. The Court found that the department was not negligent in the way in which the survey was conducted, pointing out that the tree in question had been in full leaf and apparent good health at the time of the survey.

Not all cases involve trees falling onto the road. Some involve simple claims for the costs of repair to damaged vehicles. *Kamin v. Windsor (City)* (1990) and *C.K.O.Y. Ltd. v. Ottawa (City)* (1952) are examples. In both cases the plaintiff lost on the basis that the respective authority had done what was reasonable and could not have foreseen, or in fact discovered with a routine inspection, that the tree limbs would fail.

...the plaintiff lost on the basis that the authority had done what was reasonable and could not have foreseen...that the tree limbs would fail...

Millar & Brown v. Vancouver (City) (1966) involved a truck which struck and was damaged by some overhanging branches. Initially the city was found liable, but on appeal, the judge found the city guilty of non-feasance rather than misfeasance. Based on the structure of the Vancouver City Charter, the city could not be held liable, and the action was dismissed at appeal.

A similar case is seen in *National Moving & Storage Ltd. v. Fredericton (City)* (1983). Again it involved a truck striking overhanging branches and suffering damage, with the truck owner suing the city to recover his costs. In this case the judge found that the truck driver "…was negligent in failing to keep a proper look-out and his negligence was the sole and proximate cause of the collision." The city had a well implemented strategy to ensure adequate clearance on all main truck routes, and although it could not be held to a standard of perfection, it had done what was reasonable to ensure safe passage of large trucks. In this case, the truck was not on a known truck route.

5.4 Roots Under Road Rights-of-way

In *Stockinger v. Coburg (Town)* (1944), a drain was constructed by a homeowner under a municipal road easement. The drain connected to the municipal sewer. The municipality granted the homeowner an easement to build the drain under the municipal land. The work was done at the property owner's expense but under the supervision of the municipality. The property owner sold the land to the plaintiff. The roots of a tree growing in the municipal road allowance and adjoining the land of the plaintiff grew into the drain and blocked it. The plaintiff's basement was damaged by flooding due to the blockage.

The issue in this case was whether the municipality was liable for maintenance of the drain and, when the blockage of the plaintiff's drain occurred under the municipal property, whether the municipality was responsible. The Appeal Court ruled that the drain was not part of the sewer, and could not be considered a local improvement authorized and built by the municipality. It was not part of the "soil and freehold of the highway" but continued to be the plaintiff's property. The plaintiff then, not the municipality, was responsible for the maintenance and repair of the root-clogged drain. The municipality was found not liable for the damages resulting from the flooding.

5.5 Roots Under Sidewalks

City trees are often placed next to roads and sidewalks, and it is quite common to find the sidewalk paving being lifted as the roots underneath grow in diameter and extent. Pedestrians using sidewalks with an uneven surface can trip and injure themselves. An early example is *Scott v. City of Nanaimo* (1924). A woman stubbed her toe against a block of cement sidewalk that had been lifted about 12 inches by a tree root. The Court decided that the sidewalk had been installed correctly and that the plaintiff was not entitled to damages. Her husband appealed the judgement (the woman having died of other causes in the interim). The appeal Court dismissed the appeal, albeit with a dissenting judgement in support of the appeal. The judgements in both courts hinged on provisions of the Municipal Act then in effect.

In *Hennessy v. Toronto* (1928), a contrary judgement arises. Again a pedestrian tripped over a concrete slab raised by tree roots. In this case, the city had known of the defect for some time, but because it was one of many similar defects throughout the city, had taken no action to fix it. The judge ruled that the plaintiff had a right to expect the sidewalk to be reasonably level, and especially so since this was the only similar defect in the area. The plaintiff was awarded $800 plus costs and medical expenses.

In 1953, the case of *Marshak v. Toronto (City) (1953)* comes up. The plaintiff again tripped on a raised sidewalk slab and was injured. Noting the provisions of the Municipal Act then in effect, the judge saw no duty

> to keep it [the highway] free of imperfections or as smooth as a billiard-table. These slight imperfections in the level of sidewalks are to be found in many places throughout the municipality. It would be demanding the impossible to require that they be smoothed out on pain of imposing liability for damages in case of an accident. ...If I were to impose liability in the circumstances of this case, it would be the equivalent to making the municipality an insurer of the safety of all pedestrians who suffer falls.

The action was dismissed.

A more recent case is seen in *Hugh v. Vancouver (City)* (1981). Once again a concrete sidewalk slab raised by a tree root caused the plaintiff the trip and suffer some injuries. The judge examined the state of the sidewalk, the inspection and repair programme, and the legal aspects of the Vancouver Charter. He concluded that the Vancouver Charter gives the city the power but not the duty to maintain and improve its sidewalks. The standard of care for sidewalk maintenance is decided by council as a matter of budgetary policy. It was, therefore, beyond the scope of the Court to decide whether or not the standard was reasonable. Note that this decision follows from the one made in *Anns v. Merton London Borough Council* (1978) and reflects the split between policy and implementation (See Section 1.2.3).

5.6 Tree Removal along Highways

A large sugar maple, approximately 100 years old, was growing partly on a road allowance but mostly on the land of the plaintiff in *Bottan v. Richmond Hill (Town)* (1995). The tree was inspected by the Richmond Hill urban forestry supervisor, who noted that it was declining, with dead wood, trunk rot, a poor structure and evidence of insects and the disease of false tinder rot fungus. The supervisor considered the tree hazardous to both private and public roads and also to the overhead utility lines which passed through its crown. Remedial work, such as cabling and bracing, were not viable considering the tree's age, probable remaining lifespan, poor condition, and location. As per town policy, the tree was posted with a removal notice and then cut down. The plaintiff claimed $75,000 for trespass and $25,000 in exemplary or punitive damages for the removal of the maple. In the court proceedings, an expert witness testified that in his opinion the tree was hazardous and as a consequence had no or negative value. The Court agreed that the sugar maple was a hazard tree but, before passing judgement, the Court questioned whether or not the *Municipal Act* (Ontario) provided a statutory bar to the court action as well. Section 284 (3) of the Municipal Act states:

> No action shall be brought against a corporation for the recovery of damages caused by… or on

account of any … disposition of any… tree… adjacent to or in, along or adjoining any highway or any part thereof not within the traveled portion of such highway.

Referring to this section of the *Municipal Act*, the Court dismissed the action.

In *Hall v. Etobicoke (Township)* (1963), the municipality was taken to court for the alleged wrongful removal of a 23 metre tall elm tree naturally growing on a road allowance in front of the plaintiff's property. The plaintiffs argued that they had a proprietary right to the tree even though it was not on their land. The Court ruled that the municipality could remove, without restriction and without liability to an adjacent property owner, any naturally growing tree found along a highway over which the municipality had jurisdiction. This action was also dismissed.

5.7 Summary

Trees growing along roads, streets and highways have precipitated many different court actions. In some cases, such as *Swinamer*, governments have made policy decisions which exempted them from liability. However, the government could still be held liable if implementation of the policy were found to be flawed, as in the case of *Just*. In all cases, if the inspections were properly carried out and the fault causing the tree to fail could not have been identified, even by an expert, prior to the incident, the responsible authority cannot be found liable. However, governments and other employers may be found vicariously liable for the actions of their employees in the event of an accident due to those actions. In some cases, the award made has been decreased due to the contributory negligence of the plaintiff.

Chapter 6
Trees Affecting Utility Rights-of-way and Easements

6.1 Introduction

Utility rights-of-way and easements have seen their fair share of litigation when accidents involving nearby trees have caused death, personal injury, or damage to property. The issue in the ensuing court action is generally whether the defendant, often a telephone or power company, is guilty of negligence.

6.2 Cases Involving Telephone Utilities

In *British Columbia Telephone Company v. Cooper Builders Ltd.* (1993), a 23 metre high western hemlock growing on a condominium development site fell during a windstorm. The hemlock blew onto and damaged telephone lines and three telephone poles on an easement owned by the telephone company in a school yard adjacent to the site. The hemlock looked healthy, showing no outward signs of injury, but it had root rot and had been in an advanced state of decay for several years prior to its failure. At issue in this case was whether the defendants' negligence caused the tree to fail onto the telephone lines.

The plaintiff, the telephone company, maintained that the defendants breached a duty of care by failing to assess the condition of the trees which were selected for retention on the development site, by failing to cut down any trees in poor condition, and by failing to protect the retained trees during construction of the condominium units. The judge suggested that if an arborist had done a thorough assessment of the trees, including core sampling of the trunk of the tree which failed, the hemlock would have been identified as a hazard and removed. A simple visual inspection would not have determined that the tree was suffering from root rot since there was no visible evidence. The issue then was whether or not the defendants were obliged to thoroughly inspect each tree before development of the land. The judge ruled that,. in the absence of some external evidence that the tree was hazardous and diseased, there was no obligation on the part of the defendants to have core samples taken of each of the trees marked for retention on the property. The judge further ruled that there was no evidence to suggest that the defendants should have known that the hemlock was a potential hazard. The defendant was found not liable for the damage caused by the collapse of the tree.

The plaintiff also claimed that the defendant did not take adequate tree protection measures, and that this failure contributed to further liability. The telephone company argued that the defendant should have placed a fence around the tree to prevent damage to the tree and its roots from heavy machinery. The plaintiff also argued that the existing soil grades around the tree should not have been disturbed. The Court ruled that the municipality had no such tree protection by-law in place at the time and that no statute or case law could be found to suggest that a duty to take these precautions was owed by the defendants to the plaintiff, if in fact such a duty of care did exist. The judge suggested that, even if all the tree protection measures had been ignored, there was no evidence that the lack of protection resulted in the tree's failure. The evidence conclusively showed that the tree was going to fail sooner or later, regardless of the tree protection measures used by the defendants. The Court dismissed the action for damages in negligence.

6.3 Cases Involving Power Utilities

Injuries resulting from trees and power lines connecting are documented in several cases. In *Ottawa Electric Co. v. Crepin* (1931), an eight year old boy was walking along a sidewalk when he fell over and came in contact with a live wire lying on the ground. The wire had apparently been broken by a falling tree limb. The boy lost part of his hand, including four fingers. At trial, Ottawa Electric Company was found guilty of negligence because, although the broken wire had been reported, the owner failed to rectify the problem promptly and did not implement precautions that would have protected anyone passing by the damaged wire. The case was appealed to the Ontario Supreme Court and eventually to the Supreme Court of Canada. Both courts upheld the original verdict.

In an action heard in the Supreme Court of Canada, *Amos v. New Brunswick (Electric Power Commission)* (1977), a nine year old boy was seriously injured while climbing a tree which contained a series of hidden utility lines. The incident occurred in July while the tree was in full leaf. The plaintiff and his friends were seeing who could climb the tree farthest and fastest. The plaintiff climbed high enough that his weight caused the tree trunk or one of its branches to scrape against the high voltage lines. The boy was immediately struck by severe electric shock, was knocked unconscious, and began to burn in the tree. The neighbour cut the tree down with a chainsaw; it landed on the road and the plaintiff was rushed to hospital.

The Supreme Court of Canada ruled that it is the duty of utility companies who erect power lines carrying heavy voltages to take proper precautions against the possibility of injury

The row of utility poles carrying the wires was located along the edge of a highway and directly in front of the home of the plaintiff's uncle. The tree, a fast growing poplar, was growing directly below the high voltage lines which were strung along the top of the utility poles. The first wire was a telephone cable approximately 6.4 metres above the ground. A low voltage line of 110 volts was strung 101 centimetres above this first wire, and 30 centimetres above the second one were two more low voltage wires. Above these lines was a neutral ground wire, and 1.2 metres above the neutral ground wire were the 7,200 volt lines. These last three high voltage wires were fixed to the cross bar at the top of the utility pole.

The power commission's practice was to trim trees growing adjacent to the power lines every four to seven years to prevent them from growing into the utility lines. As far as the commission's inspector knew, the poplar tree had been pruned within those timelines. The wires were at least 10 metres above the ground and contact with them could have been avoided by trimming the tree, but it had not been trimmed. The thick summer foliage hid the wires from the boy climbing the tree. The accident was foreseeable and almost inevitable given the location of the tree and the presence of children in the neighbourhood.

The Supreme Court of Canada ruled that it is the duty of utility companies who erect power lines carrying heavy voltages to take proper precautions against the possibility of injury. The Court referred to another Supreme Court decision, *Gloster v. Toronto Electric Light Company* (1906):

> The defendant company transmitting such a dangerous element as electricity through wires thus strung along a public highway fall short of being insurers, but are bound to exercise the greatest possible care and to use every possible precaution for the protection of the public.

This principle has been reiterated in many subsequent cases and was confirmed in *Moule v. New Brunswick Electric Power Commission* (1960) where Justice J. Ritchie said:

> That high voltage wires are dangerous goes without saying, and the fact that children are likely to climb trees is certainly a foreseeable circumstance and that it is one to which power companies should give heed in placing their wires so that young climbers will not come unexpectedly on a live wire concealed by the

branches of a tree is shown by the case of *Buckland v. Guildford Gas* [Light] & *Coke Co.*, [1949] 1 K.B.410.

In the case of *Amos,* the Court concluded that the tree invited small children to climb it. The Court also concluded that the power commission failed to properly prune the tree next to the power lines. The high voltage wires were a concealed danger. The defendant should have reasonably foreseen the danger and taken steps to reduce it. The New Brunswick Electric Power Commission was found negligent for damages to the plaintiff. The case was appealed, and the lower ruling overturned, whereupon it was then appealed to the Supreme Court of Canada. The Supreme Court of Canada allowed the appeal with costs and restored the original award for damages of $36,275.45

The defendant should have reasonably foreseen the danger and taken steps to reduce it

In *Neal v. West Kootenay Power Ltd.* (1993), the issue of negligence again appears. In this case a large (19.8 metre tall) Ponderosa pine tree fell across a road. The pine, which had been dead for over two years, broke approximately 1.2 metres above the ground. The prevailing wind blew the tree into a power line carrying 7,200 volts. The wire came down on a steep bank a few feet above the road. A brush fire, ignited by the live wire, put an orchard and home above the road at risk. Gordon and Christel Neal, driving home, found that the fallen pine had blocked the road not far from their driveway. Gordon Neal got out of his car and began to climb the bank in an attempt to help his next door neighbour. Mr.

Neal died instantly when his body touched the live power line which had fallen onto the bank.

The issue in this case was whether West Kootenay Power was liable for Gordon Neal's death. The power company argued that it could not have foreseen that the dead pine would collapse onto the power line, or that the fuse on the line would not burn through, de-energizing the wire before Mr. Neal contacted it. The British Columbia Court of Appeal, in an earlier case, *Leischner v. West Kootenay* (1986), said that a power company

> owes a particularly high duty of care to see that, in transmitting and distributing electricity, members of the public are not injured by the activity

Further on in the same action, the Court said that the power company

> was in breach of its duty of care in failing to take positive steps to ascertain if any danger existed, and by waiting to be informed by others of any potential danger, rather that making reasonable inquiries itself.

The dead pine was on the opposite side of the road from the utility line, but could easily be seen during any inspection of the lines. West Kootenay Power did have a tree trimming program where each area was visited every three or four years to prune or remove trees that interfered with the power lines. This program was focused toward maintaining uninterrupted service to the power company's customers and not toward public safety. Unfortunately, there was no record of any tree trimming on the road in question. In addition, it was the policy of the power company to inspect the lines annually. An inspector drove along the lines in a car and tagged each hazardous tree by shooting a marker into it. The dead pine was not marked or reported to West Kootenay Power as being hazardous by any of its employees, such as lineworkers or metre readers. In fact, there was no evidence indicating that inspections of the power lines along the road in question had occurred.

An officer of the power company, during examination for discovery prior to the trial, stated that, "where there is a standing dead tree which clearly endangers the line the company considers that tree needs to be cleared." The Court suggested that if an annual inspection had occurred, the pine tree would have been identified as a hazard by the power company, and removed.

The Court ruled that the power company had a higher duty of care than the highway department

The power company argued that since the pine tree was on the highway right-of-way, it was the highway department's duty to identify the tree as a hazard and remove it. The Court ruled that the power company had a higher duty of care than the highway department. West Kootenay Power Ltd. was found liable for Gordon Neal's death.

By contrast, the power commission was not found negligent in in *Moule v. New Brunswick Electric Power Commission* (1960). A 10½ year old boy climbed a spruce tree, and then crossed from a tree house in the spruce into a maple tree. He climbed higher and higher into the maple tree before stepping on a dead limb. The limb gave way and he fell in such a way that his body made contact with a high voltage wire and the trunk of the maple tree simultaneously. He was badly injured as the result of the electric shock. The Supreme Court of Canada ruled that the sequence of events leading to the child's injury was so fortuitous that it could not have been foreseen by a reasonable person. The Court ruled that the New Brunswick Electric Power Commission had a duty to take precautions, which they did by pruning in and around the wires, but only against any foreseeable consequence of the danger caused by the high voltage lines. In this case, the defendant was found not liable for damages.

The Supreme Court of Canada ruled that the sequence of events leading to...injury was so fortuitous that it could not have been foreseen by a reasonable person

In *MacDonald v. Goderich (Town)* (1947), a 4 ½ year old boy climbed an apple tree located on a street, managed to touch the power lines, and consequently suffered severe burns. The Trial Court found that although the tree constituted an "allurement" to small children, the defendant could not have anticipated the danger, since the height of the wires relative to the tree and the size of the uppermost branches seemed to preclude the possibility of a small boy reaching the wires.

Not all electric power cases involve injuries. *Hansen v. Nova Scotia Power Corp.* (1986) was simply an allegation of trespass. In developing a subdivision, a power pole was moved further back, and this move involved cutting down a tree which the plaintiff alleged was on his property. The trial judge dismissed the case and was upheld on appeal.

In the case of *Saskatchewan Power Corp. v. Wolf* (1992), a beaver chewed a tree causing it to fall against a power line, which triggered a fire that damaged farm buildings and the machinery and materials stored in them. The small claims court found no evidence of negligence by the Power Authority, but awarded damages for nuisance. The defendant appealed, noting that it had some rights under the defence of statutory authority under the Power Corporation Act. At appeal, the verdict was reversed because it was found that the Power Corporation was undertaking work mandatorily decreed under statutory authority, and "…unless the loss that occurred could have been foreseen or was an inevitable consequence of what was maintained by the appellant [that is, the power line itself] then it cannot be found liable for the loss which occurred."

In *Morris v. Fraser* (1965), an experienced tree faller was taking down a tree on his property. Due to decay in the centre of the tree, it did not fall in the anticipated direction. Instead, it landed close to the road and one of the branches hit a power line and broke it. Somehow the broken wire contacted the neighbour's domestic power line, resulting in a large voltage surge which started a fire in the house, causing damage valued at $1,321.00.

The central issue here was whether or not the tree faller was liable for the damages claimed—that is, was the result of the accident foreseeable in any way. As an experienced tree faller, he should have anticipated the possibility of central decay after the first undercut had been made. He was clearly aware of the danger if the tree fell onto the power line. He had taken precautions to ensure that the tree was tied to a rope and truck, and that the truck could be used to pull the tree in the direction desired. Despite these precautions, the judge ruled that the faller was negligent. However, the judge ruled that the damage resulting from the tree falling onto the power line was not a reasonably foreseeable result of the faller's negligence.

In dismissing the action, the judge cited *Sharp v. Powell* (1872) as follows:

> No doubt, one who commits a wrongful act is responsible for the ordinary consequences which are likely to result therefrom; but, generally speaking, he is not liable for damage which is not the natural or ordinary consequence of such an act, unless it be shewn that he knows, or has reasonable means of knowing, that consequences not usually resulting from the act are, by reason of some existing cause, likely to intervene so as to occasion damage to a third person. Where there is no reason to expect it, and no knowledge in the person doing the wrongful act that such a state of things exists as to render the damage probable, if injury does result to a third person it is generally considered that the wrongful act is not the proximate cause of the injury, so as to render the wrong-doer liable to an action.

6.4 Summary

The courts in Canada have ruled that public utilities, such as power companies, have a very high duty of care to protect the public from risk of injury. Liability of the defendant in cases where trees growing near utility rights-of-way have caused death, personal injury, or property damage rests upon whether the defendant was negligent and whether the events culminating in the accident could have been foreseen. For example, in several court actions where youngsters have climbed trees and inadvertently been injured by high voltage lines, the Court determined whether the events leading up to the accident were foreseeable by the power company. In cases where they were, the Court found the power company liable for damages. Where they were not, the Court ruled that the power company could not have foreseen the events leading up to the accident, and therefore was not liable for damages.

Chapter 7
Trees Damaged by Chemicals

7.1 Introduction

Pesticides (particularly herbicides and insecticides), road salt, and miscellaneous chemicals like household paint have damaged plants and given rise to several court actions in Canada. Plant injuries caused by these chemicals can range through foliage damage, twig dieback, reduction of fruit or grain production, and decline in plant vigour to plant death. Such damage has led to actions in trespass, for nuisance, and for negligence being brought before Canadian courts.

7.2 Pesticide Injury

In *Melito v. Lenbro Holdings Ltd.* (1983), a herbicide with an active ingredient of atrazine was used improperly. In this case from Ontario, the defendant hired the co-defendant contractor to regrade his property to create a gravel parking lot. As a result of the work, weeds emerged from the disturbed soil, and the contractor applied a soil sterilant to control the weeds. Unfortunately, the atrazine contained in the herbicide contaminated surface water which drained off the regraded land onto the plaintiff's property, damaging his grass. In addition, the atrazine spray drift injured a vegetable garden, newly planted fir tree seedlings, and two mature trees on the plaintiffs' property. The defendant had a duty of care to make sure that the regrading did not cause surface water to drain onto the neighbouring property. The defendant knew, or should have realized, that the herbicide might contaminate the drainage water. He should have done something to avoid damage to the plaintiffs' land. The defendant also had a duty to ensure that the work was done by a licensed pesticide applicator, which the contractor was not.

The plaintiffs were awarded $7,960 in damages by the Court. This award included $6,960 for removal of the contaminated soil by a bulldozer, replacement with new topsoil, and regrading and sodding of the grass area. In addition, $1,000 was assessed for the damage to the nursery fir seedlings. The Court also suggested that if the case had been presented differently, the assessment for damages could have been considerably higher, since no value was placed on the mature trees lost. Nor was a claim made for the loss of profits due to the death of the nursery seedlings or for the loss of fresh produce from the vegetable garden damaged by the spray drift. The defendant and the co-defendant shared liability, but the contractor was required to pay the bulk of the damages awarded because of his negligent use of the pesticide.

The defendant ...should have done something to avoid damage to the plaintiffs' land

In a second case, *Dixon v. Patton* (1986), a prairie shelterbelt of poplar trees, laurel leaf willow, spruce, mayday, green ash trees and a caragana hedge, in addition to a rye crop, were alleged to have been damaged by herbicide spray drift. In June of 1981, the plaintiff's trees were thriving and their rye crop, which was growing between the shelterbelt and the defendant's property, promised a good yield in the coming harvest. On June 22, 1981, the defendant sprayed his property with Buctril M, a herbicide containing bromoxynil and MCPA as active ingredients. The plaintiff smelled a strong herbicide odour while she was outdoors. To avoid the smell she moved into the house, where she observed the spraying of the defendant's lands from her window. The plaintiff estimated the wind speed to be 16-24 kilometres per hour in her direction, with higher gusts. The plaintiff observed dust and spray drift passing over the rye crop toward the farmhouse.

Within three days of the spray application, tomato plants growing in the plaintiff's garden were showing signs of deformity. Within three weeks, the trees in the shelterbelt were also exhibiting foliage deformation. The rye plant blooms blew off the plants and the seed heads formed later were also deformed. When the rye crop was harvested in late summer, the yield was 33% less than estimated in June. In the spring of 1982, the poplar and caragana hedge did not leaf out normally and the willows suffered twig dieback. By 1985, 93 poplar trees and all of the laurel leaf willows, green ash trees, and mayday trees (except one) had died. The caragana recovered and the spruce trees were not injured.

At issue was whether the death of the trees in the shelterbelt and the reduced yield of the rye crop were the direct result of the herbicide application on the defendant's property. A number of well qualified expert witnesses were called by both the plaintiff and the defendant. An expert in atmospheric physics and herbicide spray drift gave evidence based on the wind speed of the day in question, the height of the spray boom nozzles, the type of nozzles, the nozzle pressure, the application rate of 0.56 kilograms per hectare and the length of time the spraying took. He testified that it was physically impossible to deposit enough herbicide via drift onto the rye crop and the shelterbelt to cause the damage observed. The expert also testified that, based on the rate of application allegedly used by the defendant, the plaintiff should not have been able to smell the herbicide.

...the plaintiffs were entitled to the use of their land free from any interference from the defendant

What was the explanation for the death of the shelterbelt trees? One suggestion was that damage may have been due to the application of herbicide made by the plaintiff to control weeds around the trees several years prior to the spray drift incident. Unfortunately, this explanation does not account for the reduced rye yield immediately after the herbicide drifted onto the property. The Court looked to the testimony of the atmospheric scientist and dismissed the action. The conclusion of the Court was that it was improbable that the defendant caused the plaintiff's trees to die or caused a reduction in the rye crop that year.

An application of insecticide, rather than herbicide, was the cause of the *Friesen v. Forest Protection Ltd.* (1978) court action in New Brunswick. The plaintiffs owned a forested property which lay in the center of a larger block of land. The larger tract of land was to receive an aerial application of the insecticide fenitrothion to control spruce budworm. The plaintiff's home was located in a cleared section of the property. The plaintiffs had requested that their land not be sprayed, but while they were out harvesting fiddleheads the spraying occurred directly overhead. The plaintiffs claimed that they had been sickened by the spraying and that they had also suffered psychological stress. In addition, they claimed that bees, livestock, and fruit trees on their property had been killed or injured.

The Court ruled that the plaintiffs were entitled to the use of their land free from any interference from the defendant. As it fell or drifted onto the ground, the insecticide constituted a nuisance which caused damage. The Court considered the spraying to be a trespass by the defendant. The Court ruled that the defendant did not have statutory authority to spray private lands. Section Three of the *Forest Service Act*, (R.S.N.B. 1973, c. F-23 s. 3) did not permit pesticide application on private property to the detriment of the private rights of the property owner. The Court awarded general damages in the amount of $1,328 for medical costs, but the claim for damages for the dead bees, fruit trees, and livestock was dismissed.

7.3 Salt Injury

Schenck v. Ontario (1982) concerns damage to trees from road salt. In this case the issue was whether orchardists whose trees were injured as a result of the application of de-icing salt to an adjacent highway were entitled to recover damages from the provincial government.

The plaintiff, Schenck and his family, had owned an orchard in the Niagara peninsula outside of the City of St. Catharines for over 65 years. The orchard was split down the middle when the Queen Elizabeth Way was constructed through the land in the late 1930's. The orchard lands directly abutted the highway. Peach trees growing adjacent to the highway began to exhibit salt spray injury in the mid-1950s. Both fruit yield and fruit quality were reduced in subsequent years.

A separate action, the Rokeby case (1982), was later joined with Schenck's action because of similarities between the two cases. In the Rokeby case, the plaintiff had grown apples since 1957 in an orchard located in the County of Elgin, which abutted Queen's Highway No. 73 near Alymer, Ontario. Apple trees growing adjacent to Highway No.73 had shown symptoms of salt spray injury since 1959. In some years, trees growing in the orchard had been injured as far as 122 metres from the highway itself.

For many years both orchardists had complained to the provincial government regarding the loss of the growth, productivity, and therefore profitability of their orchards. Queen Elizabeth Way and Highway No. 73 were heavily travelled winter roads which the Ministry of Transportation had designated as bare pavement highways, meaning that they received the highest level of winter maintenance. Very large amounts of de-icing salts were applied annually. The salt eventually dissipated into the environment, some running off with the snowmelt and rainfall into surrounding waterways. Some of the salt was incorporated into the soil along the highways. Some in the slush on the highway or its shoulders was whipped up by the speeding traffic and thrown into the air, where the wind carried it into the adjacent orchards. Tree damage over the years had been extensive. Twig dieback every year caused lateral branches to become more dominant and produced a witch's broom appearance. Many of the trees adjacent to the highways were dead or had a large amount of dead wood that was non-productive.

The government of Ontario argued that the economy of the province was dependent on the safe year-round travel of its citizens on its roadways. The decision to continue to use de-icing salt, knowing that it was causing injury to fruit trees, was a reasonable one made in the bests interests of the public. The government further maintained that the economic loss sustained by the plaintiffs was unfortunate but inevitable. The Court disagreed, suggesting that to accept the government's position was in effect allowing the government to expropriate land without compensating the owners. The Court ruled that the provincial government was liable in nuisance. The plaintiffs, Schenck and Rokeby, were entitled to compensation for the loss of productivity to their orchards due to the application of road salts.

7.4 Miscellaneous Chemical Injury

Sometimes, something as commonplace as household paint can cause plant injury and become the cause of a court action. In *Yadlos v. Sturdy* (1989), the plaintiff built a 2.1 metre high cedar fence wholly inside the boundary line of his lot. The plaintiff designed, built and painted both sides of the fence without conferring with his next-door neighbour, the defendant Mr. Sturdy. The plaintiff painted his side of the fence with a clear stain but painted his neighbour's side of the fence dark brown. Unfortunately, the dark brown colour did not complement the pastel hues of the Sturdy residence. The defendant noted that Mr. Yadlos was free to build, on his own land, any fence he chose as long as it was built within the 1.8 metre height limit imposed by a city by-law. Sturdy complained to the City to have the fence height reduced to the 1.8 metre limit. He then took his self-help measures one step further and trespassed onto the plaintiff's land to repaint his side of the fence while the plaintiffs were on vacation. In doing so, the paint splattered the Yadlos' side of the fence, and damaged the bedding plants growing next to the fence. Paint spattered onto a picnic table, a wooden retaining wall, and onto the rear wheel drum of the plaintiff's van. Sturdy recognized at the time that he had caused some damage, but compounded his error by climbing over the fence and attempting to paint over the splashes he had made on the fence.

The Court awarded the plaintiff $1,739.44 to repair the damages, including the replacement cost of the dead and/or paint splattered plants. The defendant, Sturdy, counterclaimed, seeking damages for an alleged injury to his tree suffered when Yadlos carelessly applied a pesticide on his property. The Court dismissed this claim.

7.5 Summary

Plaintiffs have been awarded damages when chemical injury to their plants has been the result of an action in trespass, nuisance, and/or negligence. In one case, however, the plaintiff was unable to prove beyond a balance of probabilities that the damage had been caused by herbicide drift. It is imperative, once chemical injury is suspected, to collect evidence and call in expert witnesses to verify the damage before the chemical dissipates into the environment. Time is of the essence.

Chapter 8
Other Tree Related Court Cases

8.1 Introduction

In addition to the types of cases discussed in earlier chapters, several other tree related issues have gone to court. The cases discussed below come under the general headings of view covenants, personal injury from trees being cut down or unexpected tree failure, damage caused by trees falling across property lines, and using subdivision approvals to protect trees.

8.2 View Covenants

All three of the following cases originate in British Columbia, but may be applicable elsewhere in Canada. Each of these cases revolves around a covenant that had been registered on subdivision lands when the subdivision was created. This covenant prevents any owner within the subdivision from having vegetation that restricts views.

The case of *Purdy v. Kneip* was heard in 1975. The case involved a subdivision in an area known as Marlborough Heights in North Vancouver. A covenant registered on the subdivision stated:

> 10. No trees, shrubs or other growth shall be allowed to grow up or remain on the premises in such a way or in such a place or places as shall or may interfere with the view of other residents of the said Subdivision.

The plaintiffs argued that there were several trees that blocked their views and contravened the covenant. It was shown in evidence that the trees in question were fully or partly grown and obviously in existence within the area affected by the covenant before the subdivision was initiated. The judge was unable to find any cases where a court of equity had ordered a landowner to correct a problem that neither he nor his predecessors had created. The judge ruled that the covenant only applied to vegetation that came into being after the subdivision was built. Additionally, the judge found that removal of the largest and oldest tree would adversely affect both property values. Consequently, a mandatory injunction applied in this one case could adversely affect the value of the many other properties also subject to the covenant. Since the Court did not know the exact nature and extent of these potentially adverse effects, the judge declined to apply such an injunction.

...ruled that the covenant only applied to vegetation that came into being after the subdivision was built

In the case of *Cloutier v. Ball*, which was heard in 1995, the covenant was registered in the 1950's and stated:

> 10. No trees, shrubs or other growth shall be provided to grow, be, or remain on the premises in such a way, or in such a place or places as shall or may interfere with any poles or wires erected for electrical or telephone purposes in the right-of-way immediately adjoining the said premises or interfere with any guy wires to support such pole line which may be placed and maintained on the said premises or with the

> view or access to light and air from other lots in the said sub-division and without limiting the generality of the foregoing, such trees or shrubs shall in no event exceed 20 feet in height.

In the intervening years many trees had grown taller than the stipulated 20 foot maximum, there had not been any complaints in court, and enforcement of the covenant had been irregular. Mr. Cloutier and Ms. Briscoe launched a suit against Mr. Ball and several other people in an attempt to have their views restored in accordance with the covenant. The action failed for several reasons, but primarily because it was shown that compliance with the covenant by the defendants would only partly restore the plaintiffs' view. For the plaintiff to have complete views, as outlined by the covenant, several additional property owners would be required to comply, and they had not been named in the action.

The action failed...primarily because...compliance with the covenant by the defendants would only partly restore the plaintiffs' view

Moreover, the plaintiffs had not maintained the view they originally had in 1966 when they purchased their lot. In the interim, hundreds of trees had grown up which affected their view, yet they waited over a decade and a half before taking any action. Issues of equity were clearly involved. Court ordered compliance against the defendants would be inequitable, since the plaintiff's view would still be obscured by trees on other properties, and these people were not cited in the action. The covenant in its original form was a negative covenant; that is, it specified that certain things, such as growing vegetation beyond twenty feet tall, were not to happen. Requiring compliance decades later would change it into a positive covenant because considerable amounts of money would have to be spent to restore the views if strict compliance were effected. Under the law, a positive covenant cannot be made to "run with the land." As a result, any subsequent owner of any parcel of land affected by the covenant could not be forced into compliance because the people entitled to enjoy views had not sought compliance from the previous owner.

The defendants filed a counter claim seeking to have the original covenant declared obsolete, on the basis that the neighbourhood character had changed sufficiently for the covenant to now have no meaning. This too was denied by the judge because there was some evidence of compliance in the overall area, and in fact, some residents may have traditionally relied on it as a means of retaining their views. Again, issues of equity would arise by cancelling the whole covenant.

Finally, the issue of views and covenants arises again in an ongoing case: *Bosch v. Hudson*. This case is also in the Marlborough Heights area of North Vancouver. The plaintiff, Mrs. Bosch, seeks to have the covenant applied so that her view will be unimpeded by trees located downslope on the property of Mr. and Mrs. Hudson. At the time of writing this case was not completed.

8.3 Personal Injury from Trees Cut Down

Injuries resulting from falling trees are common in industrial forestry. There are also several cases where people have been injured as a result of negligence by the tree fallers. In *Dykeman (Guardian ad litem of) v. Emerson* (1991), a small girl (Amanda Dykeman) aged three, was trapped by a tree cut down on a neighbouring lot owned by the Emersons. Their contractors were clearing trees for firewood and had done so previously on numerous occasions. On the day of the accident, they had already cut down three trees, and each time, one person had kept a lookout for the children playing next door, while the other cut the tree down. The lookout was not maintained for the fourth tree, and the parent (Mrs. Dykeman), momentarily occupied in the house, did not see her children wander towards the cutting area. The fourth tree fell across the lot lines and trapped Amanda (no record of injury is noted). The plaintiffs accused the two workers of failure to warn the neighbours, and failure to keep a proper lookout. They accused the land owners, the Emersons, with a breach of their duty of care to keep their premises reasonably safe, with employing unqualified workers, and with a failure to effectively supervise these workers. The defendants responded by denying a breach of their duty of care and claimed that Mrs. Dykeman was wholly or partly at fault because she had not properly looked after her child (contributory negligence).

The judge found that the two tree fallers did fail to exercise a duty of care towards the child, and said:

> *"Tree falling, an inherently dangerous activity, brings with it a high standard of care..."*

> Tree falling, an inherently dangerous activity, brings with it a high standard of care and obliges those engaged in it that they keep an unobstructed view where the tree is going to fall and ensure that no one comes within the zone of danger.

The defendants' claim that the accident was not foreseeable was dismissed, since they had foreseen the possibility for the first three trees, hence the use of a lookout. Mrs. Dykeman was exonerated on the basis that any normal parent could not be expected to watch over her child 100% of the time; she had warned the children, and had tried to keep them within her sight most of the time. The case against the Emersons was dismissed because they had reason to believe that the fallers could be relied upon to work safely, and had no reason to warn the Dykemans every time a tree was to be removed. However, their costs were awarded against the tree fallers, rather than the plaintiff.

In *German v. Welyki* (1984), two friends were cutting firewood in a poplar grove containing trees 30 to 40 feet tall. The two were cutting trees down, working about 15 to 20 feet apart. The defendant was in the process of falling another tree and checked to see where the plaintiff was. All seemed well so the defendant proceeded to notch the tree and then fall it. As the tree started to fall he realised that it was going to fall about 45° away from where he intended, and he then noticed the plaintiff had moved and was kneeling down while bucking up a previously fallen tree. The falling tree hit the plaintiff, who suffered severe injuries, including damaged vertebrae and lacerations. The plaintiff sued the defendant for damage sustained.

The judge found that both parties were negligent. No discussion had taken place about safety, but because the defendant admitted he had looked to check his partner's location prior to falling the tree, he bore 75% of the responsibility. The plaintiff was also negligent. He knew his partner was falling trees but made no effort to ascertain where he was and if he was likely to be hit. He was found to be 25% responsible (contributory negligence). The judge awarded the general damages at $20,000 with special damages at $12, 079, both figures being apportioned 75% against the defendant.

8.4 Personal Injury from Unexpected Tree Failure

Claims for damages arising from tree failures are well documented in Canada and the United Kingdom. In the earlier decisions, such as *Caminer and Another v. Northern and London Investment Trust Ltd.* (1950), the owners of defective trees were found not liable for damage arising from failure when it could be clearly shown that the defective tree part was not visible, and could not have been detected by a reasonable layperson. In *Reed v. Smith* (1914), a large and decayed cedar tree fell down in a strong wind and damaged a neighbouring house. The tree was "in a state of nature," meaning it was a wild tree remaining after the land had been subdivided, and the defendant had done "nothing either to cause it to fall, or to prevent it from falling…." Initially the defendant was found negligent, but on appeal this finding was overturned.

In *Patterson v. N. Vancouver Board of School Trustees (1928),* a large and very obviously decayed tree, situated on private lands adjacent to a school, fell and seriously injured a schoolboy walking home. The Board of Trustees was sued, as well as the owner. The Court of Appeal found neither defendant liable for the damages. The Trustees were not liable and had no duty of care to ensure safety beyond the boundaries of the school premises. The landowner was not liable (by a vote of two to one) because, at that time, it was felt that an owner did not have a duty to protect the highway from his trees. Since *Patterson*, the onus on owners of trees has shifted, and if it can be shown that the defects were (are) obvious, and clearly visible to a reasonable person, liability may ensue in the event of failure and consequent damage.

...if it can be shown that the defects were (are) obvious, liability may ensue...

The *Swinamer* case in Nova Scotia and the *Just* case in British Columbia, both of which are discussed in the context of highways in Section 5.3, are high profile examples of injury due to tree failure, and both cases went to the Supreme Court of Canada for resolution. There are several other recent cases of interest.

Hayes v. Davis (1991) is a particularly important case that has been referenced in several subsequent judgements (also see discussion in Section 1.2). On October 24th 1986, Mrs. Hayes suffered injuries when two trees growing on her neighbour's property blew down, falling across the property line onto her. Both she and her husband had previously, on several occasions, mentioned to the neighbour, Mr. Davis, that they were concerned about these trees and felt threatened by them. Mr. Davis acknowledged that he had spoken to the Hayes' about the trees but had declined to have them removed on the grounds that it would cost him money. Additionally, Mr. Hayes claimed to have discussed other trees in the area that had blown down or bent over so much that he was concerned they would uproot and fall on his house; Mr. Davis could not recall this conversation.

The trial judge applied the principles set out in *Leakey v. Nat. Trust for Places of Historic Interest or Natural Beauty* (1980), which are summarised in that case as follows:

> A person on whose land a hazard naturally occurred, whether in the soil itself or in something growing on the land, and which encroached or threatened to encroach onto another's land thereby causing or threatening to cause damage, was under a duty, if he knew or ought to have known of the risk of encroachment, to do what was reasonable in all the circumstances to prevent or minimise the risk of the known or foreseeable damage or injury to the other person or his property....

Accordingly, the judge for *Hayes* noted:

> I find the defendant liable for damages ... [he] was aware of the possibility and potential threat from previous experience and discussions with

> the plaintiff's husband of trees falling or being uprooted in the neighbourhood and adjacent properties from windstorms. …this particular cluster of trees from which the two trees fell during a high windstorm striking the plaintiff were specifically brought to the attention of the defendant in the late spring of 1986, some five to six months before the accident of November 1986, as constituting a sufficient hazard to warrant their removal.
>
> The parties in the development of their respective properties had removed numerous trees and the defendant by leaving his cluster of trees under the prevailing circumstances in 1986, with previous experience of trees frequently falling during wind storms in the immediate area and made aware of the real and not imagined concerns of his neighbours, the plaintiff and her husband regarding the cluster of trees knew, or ought to have appreciated, that these trees constituted a hazard and risk to his neighbours which he permitted to continue. The defendant had a duty to take reasonable steps in all the circumstances to abate or minimize the known or foreseeable risk or damage or injury because of the hazard to the plaintiff or her property, but negligently failed to do so even though the cost was minimal….

Mr. Davis appealed the decision. His lawyer argued that the trial judge had erred in his interpretation of *Leakey.* It was acknowledged

> That an occupier is under a duty to take reasonable steps to prevent a hazard on his property from encroaching and causing damage where the occupier is, or should be, aware of the hazard.

However, it was also argued that merely expressing concern about the trees, which by all accounts appeared to externally healthy, was not sufficient, by itself, to prove that the trees were hazards, and consequently, there could be no liability attached to Mr. Davis.

The crux of this case hinges on the tests of "the reasonable man" and reasonable foresight. The appeal was dismissed, albeit with a dissenting judgement. At appeal, the majority verdict was that Mr. Davis had been sufficiently notified by Mr. and Mrs. Hayes about the potential for his trees to fall over and cause damage, either structural or personal, to the Hayes. Mr. Davis chose to do nothing, and therefore did not take reasonable precautions to abate a hazard that was foreseeable. Justice McEachern, in his minority dissenting opinion, stated that:

The crux of this case hinges on the tests of "the reasonable man" and reasonable foresight

> Liability cannot be imposed upon a landowner for damage caused by a falling, healthy tree merely by the expression of concern when there are no facts known to either party sufficient to establish liability.

He went on to note:

> If there is liability in this case, there would be liability for any healthy tree that has been the subject of expressed concern and later falls and causes damage, such as to a passing pedestrian, or motorist. That would be a general liability for any risk that may arise after any expression of concern even when there is no reason to anticipate risk beyond obvious possibility.

In *MacKenzie v. Vancouver Lawn Tennis and Badminton Club* (1978), a woman was entering the premises of the Vancouver Lawn Tennis and Badminton Club on the same day as the club was having some trees pruned in the car park area. The club manager had asked about safety of the club's patrons. The contractor had assured the club that they would not permit the public into the area where the pruning was to be done. The club, for its part, posted a cardboard sign attached to a rope across the entrance area, and made sure that there were no cars parked in the park area. The plaintiff drove to the club, saw the yellow rope and a sign saying "no parking" and then parked nearby. She proceeded to walk across the car park, and encountered one of the ground crew who bid her "Good Morning." As she proceeded across the car park a large branch was cut from high up in one tree. It fell, striking her on the head.

...the plaintiff bore some responsibility for her injury

Earlier, another club member had a similar experience when a branch fell close to him, missing but frightening him sufficiently for him to tell the crew to "be more careful." At trial there was contradictory evidence about the sign, whether it was in place and what it actually said. There was also confusion as to whether or not the ground crew had properly warned club members. On balance, the judge decided they had not maintained a proper lookout. The contractor claimed the woman had been asked to wait while work proceeded, but she denied this. She also claimed not to have heard the noise of the chainsaw up in the tree, but this claim was not accepted.

The verdict was that the plaintiff bore some responsibility for her injury. There was a sign noting that the parking area was closed for tree work, and she had heard the chain saw, but failed to heed the warnings that these offered. The contractor was also negligent. They owed the plaintiff a duty of care but failed to carry it out. The club was negligent because merely posting a sign on a rope was not sufficient warning. The contractor and the club were found liable jointly for two-thirds of the damages. Since the club had been assured by the contractor that they would look after safety issues, the club was entitled to succeed with its third party claim against the contractor. As a result, the plaintiff was awarded $8,000 in damages to offset her pain and suffering.

Another recent example is found in *Kelley v. Ingersoll (Town)* (1996). The plaintiff, Mr. Kelley, was playing bowls at the Ingersoll Lawn Bowling Club when a large limb from a white ash fell on him, causing severe injuries. Prior to the court hearing, general and special damages and prejudgement interest were settled at $125,000 plus the plaintiff's costs of $21,328. At issue was who was liable for these damages and costs: the Municipality of the Town of Ingersoll or the owners of the tree, Mr. and Mrs. Vint.

In determining liability, the judge examined the town's approach to municipal trees

Evidence revealed that the tree was mainly on the Vint's property, although it encroached onto the municipal road allowance by between 3.8 and 6.3 centimetres. After the limb failed, it became clear that the ash tree was in serious decline, with much of the core rotted away. The town denied responsibility for the tree, so the Vints had it cut down. In determining liability, the judge examined the town's approach to municipal trees. It had a tree policy in effect at the time of the accident, but it became clear that the inspection procedures were neither well implemented nor of a standard that might be expected. Mr. Vint, under cross examination, stated that he always felt the tree to be healthy, as judged by the presence of leaves, although he had no experience in dealing with trees. An expert was called in and he noted that, based on the evidence seen by him, he would have declared the tree a hazard and recommended its removal. This evidence included clearly visible defects and old wounds in the trunk which had been filled with concrete. The expert felt that an inspector would have noted the hazardous condition of the tree, even on a casual drive by, while the layperson may well not have appreciated what the evidence on the tree trunk meant.

Because the tree that shed a limb qualified as a tree on public property (due to its slight but undeniable encroachment on the road allowance), the judge found that the town should have included it in its inspection procedures. However, the town made no effort to locate and inspect trees on public property in a systematic manner, and therefore failed in its duty to protect the public. Moreover, the person responsible for inspections was not trained to recognize hazard trees, and his inspections were "unscheduled and cursory." As a result the Municipality was negligent and their negligence "resulted in a failure to identify the danger represented by the tree in this accident and caused or contributed to the damages suffered by the plaintiff."[1]

However, the tree was mainly on the Vint property and the judge found that, given its age and the condition noted by the expert,

> [a] reasonable property owner would have inspected the tree to ascertain whether it needed attention. Such an inspection would have led Mr. and Mrs. Vint to the conclusion that the tree should have been examined by a tree service. As a result, I find that Mr. and Mrs. Vint were negligent in failing to inspect the tree, in failing to observe its condition and in falling [*sic*] to take steps to eliminate the hazard represented by the tree … their negligence caused or contributed to the injuries and damages suffered by the plaintiff.

The judge apportioned liability at 65% for the Vints and 35% for the municipality. Note the difference in this case from the arguments made in *Bottoni v. Henderson* (1978) and *Doucette v. Parent* (1966), discussed in the next section. In the case of *Kelley,* the Vints were found liable because, although they were lay persons, the external evidence on their ash tree, as noted by the expert, was manifestly sufficient to cause the "reasonable person" to investigate the health and safety of the tree in more detail.

Finally, a personal injury case involving a woman tripping over a guy wire attached to a newly planted street tree is seen in *Miller v. Kamloops (City)* (1996). Initially, the plaintiff sued the City of Kamloops, the consulting company that wrote the planting specifications, and the contractor that installed the tree. Subsequently, the contractor was found to have no part in the case. This case was still before the courts at the time of writing.

8.5 Damage from Trees Falling across Property Lines

Trees commonly fall across property lines, causing various forms of damage. Some examples were discussed in the previous section, and the case of *B.C. Telephone Company v. Cooper Builders Ltd.* (1993) (in which a hemlock on one side of a fence fell down and damaged the telephone lines and poles on the other side) was covered in Section 6.2.

In *Bottoni v. Henderson* (1978), a large white pine growing on the Henderson's property snapped in a windstorm and fell onto the roof of the Bottoni's house, damaging the structure as well as some of the interior. The plaintiffs claimed that the Hendersons were negligent in not looking after their tree, and further, that the tree was a nuisance that subsequently damaged their house.

In evidence it was shown that the pine tree, located about a foot and a half into the Henderson property, snapped at about 35 feet. At the point of failure the tree had advanced decay. Prior to the incident the plaintiffs had enjoyed the shade that the tree provided, and the tree had not been an issue between the two parties. Experts called to assist the Court agreed that there were no external indicators of decay on the tree that should have been noticed by the defendants. As a result, the judge ruled that, in the absence of any reason to suspect decay, the defendants were not negligent for not cutting the tree down or taking steps to maintain it in good health. They honestly thought it was in good health and had no reason to suspect otherwise.

With regard to the claim for damage resulting from nuisance the judge examined the evidence and found that, although the branches of the pine clearly extended well into the Bottoni property, the Bottonis had not previously complained of their presence. Had the damage to their house resulted from the overhanging branches alone, the plaintiff would have had a good basis for claim. However, the damage was caused by the weight of the falling tree trunk, and since the tree trunk was on the defendant's property, it could not form part of the claim in nuisance. Consequently, the defendants were not strictly liable. The judge went on to note:

> I am of the opinion that there is no liability for nuisance where the use of property is a natural use such as the growing of a tree. I am of the opinion in this case that the maintenance of the tree by the defendants on their lands was a reasonable use of the lands so far as the plaintiffs were concerned.

The action was dismissed.

A similar issue arises in *Doucette v. Parent* (1966). Again, a large tree snapped (at about 1½ metres above the ground in this case) in a windstorm and damaged adjacent property. As with Bottoni, experts testified that the tree seemed to be externally sound with no history of external problems. The tree was believed to be a Lombardy poplar that had severe internal decay. There was some conflicting evidence about the condition of the tree, but ultimately, the judge accepted that it had looked fine before the accident. In the absence of any indications that might lead to concern, the judge noted:

> A property owner must take steps to ensure that trees located on his/her property are not hazardous to others. It would seem to me that a property owner can be held liable in negligence for failing to maintain a tree in circumstances where the exterior of the tree provides sufficient warning signals. The warning signals would include general deterioration, a scarred trunk, discoloured bark, and a large amount of defoliation or dead branches. Liability in negligence should be found against an owner of a tree in circumstances where a reasonable person, on viewing or inspecting the exterior of the tree, would conclude that the tree is in need of expert attention.

Lacking any evidence of this kind, the defendant could not have been expected to know that the tree was a danger, and therefore was not negligent or liable for the damage that resulted. The issue arose, as in Bottoni, for a claim in nuisance. The judge referred back to three main lines of authority about the appropriate tests of nuisance, and concluded that none of the tests were met in this case. The implications of *Hayes*, already discussed earlier in this chapter, were examined by the judge. He concluded:

> With respect to latent defects, a defendant is not liable in nuisance unless he/she fails to remedy it without undue delay when he/she becomes aware of it or with ordinary and reasonable care should become aware of it.

The claim by Doucette was therefore dismissed.

8.6 Using Subdivision Approvals to Protect Trees

The case of *Kahl v. Oak Bay (District)* (1978) provides an interesting perspective on the use of legislation to prevent certain actions. In this case, Mr. and Mrs. Kahl, the owners of a 36.6 metre (120 foot) wide parcel of land, applied to have it subdivided into two 18.3 metre (60 foot) wide lots. Their application was duly published and comments solicited from neighbours, as legally required. A majority of the neighbours objected, noting that the trees on the existing lot gave to the neighbourhood character that would be lost by cutting down enough trees for a house to be built on each lot. As a result of the petitions received, the approving officer rejected the application, noting that:

> It is considered that dividing such a lot into two 60 foot wide lots such as you propose would anticipate development and so remove trees that would injuriously affect established amenities of adjoining or adjacent properties.

The Kahls disagreed with this decision and launched an action to have it overturned. The plaintiffs' lawyer argued that trees on a petitioner's lot cannot be regarded as an established amenity, since regardless of whether the owners intend to develop, they have the right to cut down the trees at any time. Consequently, the approving officer cannot anticipate removal of trees solely as a result of development. The judge disagreed, noting that the construction of two new houses would certainly lead to loss of trees, and therefore the approving officer was correct in

> ...regarding the trees as an established amenity even if they were on private property. ...The approving officer is entitled to proceed on the assumption that if approval is withheld the trees will be spared, and amenities thus preserved. ... There is no suggestion that he did not act in good faith, or that he discriminated against the petitioners. It cannot be said that there was no evidence to support his decision, and there is no basis for saying he erred in law.

The action was dismissed. The case is interesting in that trees on private land were seen as public property in the form of "an established amenity" which could not then be removed at the whim of the owners. This case is a bit of an anomaly in that municipal bylaws are ordinarily used to protect trees, as discussed in Chapter 9.

8.7 Summary

The miscellaneous cases in this chapter yield a variety of judgements. View covenants, for example, may be difficult to enforce if the vegetation in question predated the subdivision or if enforcement over the years has been irregular. An action to enforce a view covenant should name all of the parties whose vegetation obstructs the plaintiff's view. In personal injury cases, plaintiffs may be found guilty of contributory negligence if they failed to take sufficient care. The prevailing ruling in cases of property damage due to tree failure is that, if the tree was visibly defective to a lay person, there may be a duty of care attached to the owner of the defective tree. If, however, the tree bore no obvious defects, the owner may not be found liable. In the last and perhaps most anomalous of the miscellaneous cases, trees on private land were seen as an established public amenity.

Chapter 9
Tree Preservation By-laws

9.1 Introduction

The courts may be bound in their analysis not only by precedents like those cited in previous chapters, but also by federal, provincial, or municipal legislation. For this reason, some precedents may be applicable only in certain provinces, especially where the precedents involve highways legislation or requirements governing the activities of provincial or municipal staff.

Tree cutting, in particular, is often regulated by municipal or provincial legislation in Canada. Typically, there is a specific piece of provincial legislation, such as the Trees Act in Ontario, or individual municipalities may make their own by-laws following enabling sections in the relevant provincial legislation, usually the Municipal Act.[2] Similar arrangements are found in other provinces, each with their own particular nuances.

Tree cutting by-laws have proven to be extremely controversial in British Columbia. The British Columbia Municipal Act was amended in the early 1990s to permit municipalities to create their own tree cutting by-laws under sections 929.01 through 929.07 (Municipal Act, R.S.B.C. 1996). The use of tree cutting by-laws to prevent development is specifically prohibited unless compensation or alternative development options can be agreed upon (see Section 9.3).

9.2 Creating Successful Tree By-laws

Publications such as those by Bernhardt and Swiecki (1991) and Hoefer and De Voto (1990) have useful discussions about crafting effective tree by-laws, or ordinances, as they are termed in the United States. The following is a synopsis of some of the key points.

To be legally binding, the by-law must be enabled by legislation, either a specific act such as the Trees Act in Ontario or more general legislation, such as the Municipal Act in British Columbia. The by-law must conform to other legal requirements as well. For example, any offenses claimed under the by-law must be administered in a specific manner. If there is an Offense Act in the jurisdiction, then care must be taken to align its requirements with those in the by-law. In most jurisdictions, some pieces of legislation will have greater authority than others. If one piece of legislation stipulates a maximum fine amount, it is often possible to increase this maximum by specifically cross referencing the by-law to more powerful legislation.

The by-law must include, as a minimum, the following specifically worded items, which will vary to some extent by jurisdiction.

1. State the purpose of the by-law and its enabling legislative authority.
2. Carefully define the technical terms to be used. For example, define what is meant by a "qualified person," which may be, for example, an ISA Certified Arborist, a Landscape Architect registered in the province, or

an American Society of Consulting Arborists Registered Consulting Arborist.

3. For any tree diameter measurements that may be required, state the height above ground where the measurement is to be made. Note that the US standard is 4 feet or 1.4 metres above ground. The international forestry standard is 1.3 metres above the point (that is, the exact place) of germination. The difference is small but could amount to a difference of many dollars if a high taper were found on the trunk. Most provinces have some form of nursery standards. In British Columbia, there is a set of Landscape Standards jointly prepared by the Nursery Trades Association and the Landscape Architects. If such standards exist and contain technical aspects that have merit in a by-law, they should be specified or at the very least referred to as the base standard of performance.
4. Note what exceptions to the by-law are permitted. Exceptions might include higher statutory legislation such as is often seen for the power lines of utility companies or for highways authorities if the tree is within a road easement. Note that is common for registered land surveyors to have the right of access to private land without consent, and they also have the right to cut vegetation as required for them to do their survey. If necessary, include a map at the back of the by-law to delineate which areas of land are included.
5. Note who has authority to enforce the by-law, and under what circumstances these

enforcement officers may enter upon private property (usually for emergency work only or to notify the owner of a problem).

6. If a list of tree species is included, the species must be clearly specified with common and scientific names. Similarly, if used, size criteria must be specified clearly. If special trees, such as heritage trees or the largest trees of each species, within the areas affected by the by-law are to be listed, then the general terms are specified within the body of the by-law and the list of trees and locations is appended. In such cases, a clause should be inserted to the effect that the list appended may not be current, and applicants must first check at City Hall for the most current list.
7. For those wishing to remove trees covered by the by-law, the permit application process must be spelled out clearly, along with all of the requirements to be met. Information required from the applicant might include: name of applicant; address of applicant and property; number, species, and diameters of trees affected; scale map showing tree locations; reasons for wanting to prune or remove trees; method of replanting; and if applicable, any other environmental conditions that might apply. Information provided to the applicant as part of the application should include a copy of the tree by-law and references to other relevant legislation. For example, in British Columbia the *Wildlife Act* specifically prohibits cutting trees in which certain birds, such as eagles, ospreys, and blue herons, are nesting. Drawing the

applicants' attention to such issues places the onus on the applicant, who cannot subsequently claim ignorance of the law. Nor should the applicant be able to claim ignorance of the by-law. To eliminate this possibility, the permit application should contain words to the following effect, "I, ________, have read the City of ________ Tree Retention By-law and agree to abide by its contents," to which the applicant must sign upon completing the application.

8. In certain circumstances, the municipal authority may want to require work on private land to abate a potential hazard to the public. Provision must then be made to notify the owner properly (which may link to other notification procedures and Acts), requesting that the work be completed within a certain time frame. If the work is not so completed, the authority may do the work itself and recover the costs later, usually by adding them to the property tax bill. Such provisions must be carefully worded if they are to be enacted successfully.

...the by-law must have political support and be practical

Finally, the by-law must have political support and be practical. There is nothing worse than having a by-law that is either so complex that it cannot possibly work in reality or so simple that most situations do not apply. In either case, the result is ineffectiveness and wasted staff time and money. It is also important to allow flexibility in by-law application while ensuring a degree of equity.

Once the by-law is in place, staff and politicians must be prepared to enforce it if it is to have any utility. In the past, enforcement has often been hampered by poor evidence. It is therefore very important for municipal staff and others gathering evidence for possible prosecution of a case to be thorough in their investigations. Specific practices (see Section 10.6 for some examples) must be followed to ensure that improper gathering of evidence or some other slip-up does not lead to dismissal of a case. Westley (1995) provides an example in which a well prepared case was thrown out because the investigating officer was not properly sworn and appointed under the prevailing legislation.

Tree protection by-laws are often controversial and do generate additional work loads for municipal staff.

9.3 By-law Challenges

Using by-laws as a means of preventing certain actions is legitimate, but using them as a means of preventing development can cause problems. Municipal tree retention by-laws have been successfully challenged for a variety of reasons. The case of *Valentine Lands Corp. v. Gibsons (Town)* (1992) provides a good example of what can happen.

The plaintiff, Valentine Lands, owned a block of land consisting of three lots made up of an oceanfront, a bluff, and a plateau area. All three parts had been used by the local citizens for many years, especially the waterfront area and its local landmark, known as Gospel Rock.

Over time there had been considerable public opposition to any development and tree removal within the general area. In 1985, as a part of the Official Community Plan (O.C.P.), the bluff and oceanfront areas were designated as sensitive to development. In 1988, a rewrite of the O.C.P. designated the bluff and oceanfront areas as a development permit area. In 1989, By-law 609 was used to require a permit for any new tree cutting in these two areas. The applicant wished to develop the lots for residential housing and, as part of the development application, agreed in 1990 to dedicate part of the bluff and the whole oceanfront area as park land, in exchange for rezoning of the plateau area. The Gibsons City Council adopted By-law 623 in September 1990, which designated the oceanfront and bluff areas as tree-cutting prohibited areas. However, in the same month, the Council refused to rezone the plateau area, which in 1991 became a tree-cutting permit area under By-law 623-1. The Municipal Act in effect at that time specified that areas could be designated as tree cutting permit areas if they "may be subject to flooding, erosion, land slip or avalanche…" The plaintiff charged that there was insufficient evidence to justify applying the tree cutting by-laws to any of the three areas and that, in effect, the by-laws were being improperly applied to prohibit development.

...using bylaws as a means of preventing development can cause problems...

This case hinged on whether the subject lands were vulnerable to the hazards identified, and hence, whether the Council was justified in introducing the by-laws it did. The opposing sides presented conflicting engineering and geotechnical reports. During the many public hearings, there was vociferous public opposition to development and tree cutting. The plaintiff also charged that the by-laws were enacted in bad faith, more to placate public opposition than because of technical merit.

The judge found that the bluff and oceanfront area had long been designated as an area that the town would like to have as parkland, and that this desire and the initial tree cutting permit by-law predated Valentine's interest in the area. Therefore, there could be no question of bad faith. That the by-law subsequently had a negative economic impact on Valentine's plans was not relevant. However, the judge also found that the plateau area was designated in response more to public sentiment than any clear evidence that development would cause irreparable environmental damage. As a result, By-law 623-1, which designated the plateau area as a tree-cutting permit area, was quashed. It was noted that the original rezoning application had been based on a compromise in which the bluff and oceanfront areas would be designated in exchange for a rezoning of the plateau area. This option remained open to the Council after the judges decision.

Another case is that of *Whistler Housing Corp. v. Whistler (Resort Municipality)* (1994). The municipality became aware that the plaintiff was planning to cut down trees on its 21 acre lot. In response, the municipality processed a new by-law, pursuant to section

929.01(1) of the Municipal Act. The new by-law expressly restricted the cutting of trees without a permit, and required the applicant to have obtained a building permit before the tree cutting permit would be issued. The plaintiff argued that the new by-law was, among other things, discriminatory and unreasonable. In reviewing the case, the judge found that the Municipal Act did grant the municipality the power to enact the by-law, as well as to apply the by-law as enacted. On this basis the petition to have the by-law quashed was dismissed.

By contrast, *Willard G. Hallman Lumber Ltd. v. Grey (County)* (1985) resulted in a different outcome. The County of Grey had passed a by-law pursuant to the Trees Act of Ontario. Section 5 of the by-law specifically required a permit to be issued before any cutting could take place. However, the Tees Act did not give any authority for this provision and it was found to be invalid. Section 5 of the by-law was therefore quashed.

...the Municipal Act did grant the power to enact the by-law, as well as to apply the by-law as enacted

In *Corporation Municipale D'Ulverton v. Brock et al* (1983), the defendants purchased the right to cut wood on a 75 acre parcel of land. In response to a petition signed by 90 residents, the municipality adopted a by-law to prohibit all tree cutting other than for farming purposes. The case revolved around whether the by-law was valid and whether it could be applied retroactively to the defendants. The judge ruled that although the by-law was valid, it could not be retroactively applied, and that the defendants had the right to continue to log the remainder of their land.

9.4 The Applicable Legislation

In many of the cases cited, and especially where municipal government is involved, the issue of applicable legislation may arise. Most municipalities claim some degree of indemnity from certain forms of accident, no matter how it was caused. The provisions of the relevant Municipal Act or provincial highways legislation are commonly cited as means of avoiding liability. See, for example, *Just*, *Swinamer*, *Schenck*, *Mendez*, and *Young v. Toronto (City)* (1985). All these cases involve interpretations of the legislation and the extent to which it applies in any one case. Because these interpretations are constantly being tested, it can be expected that new variations will arise. The precedents are an important guide to how a claim might be framed and its likelihood of success.

9.5 Summary

Tree protection by-laws are commonly used across Canada. To be effective they must be carefully written, so that both applicants and enforcers can make the by-law work in the field. By-laws must be cross-referenced to other relevant legislation and can only use the powers specifically granted in that legislation.

Chapter 10
Advice for the Arborist Acting as a Consulting Expert or an Expert Witness

10.1 Introduction

It is important to clearly understand the difference between the consulting expert and the expert witness, though a high standard of professionalism is required in both roles. As a consulting expert, the arborist can prepare one or more reports for the lawyer, discuss many aspects of the case, and be an advocate for that side of the case. As an expert witness, the arborist must be completely impartial and cannot be an advocate at all. Consequently, an arborist who has worked on a case as a consulting expert may be in conflict of interest if he or she later becomes an expert witness in the same case.

A crucial part of all legal cases involves gathering and assessing evidence. In this chapter, the basics of assembling and presenting usable evidence are outlined. The requirements of an expert report (Section 10.7) are discussed, as well as the characteristics of a successful expert witness (Section 10.9). Readers seeking more detailed information should consult Feder (1993), one of the three texts that are considered essential reading for the consulting arborist. The other two are Abeyta (1995) and the **Standards of**

Professional Practice adopted by the American Society of Consulting Arborists (ASCA) in 1996.

Reference is made in this chapter to provisions of the B.C. *Evidence Act* and to the B.C. *Supreme Court Rules* dealing with expert evidence. Similar legislation is in force in most other provinces and, at the federal level, in the *Canada Evidence Act* and the *Federal Court Rules*. In all cases, the consulting arborist should expect that the lawyer will know these provisions and explain what can be done.

10.2 What is Evidence?

To be an effective consultant or expert witness, the arborist needs to understand the basic rules of evidence and the meaning of several key words.

A fact is defined in Black (1990) as:

> A circumstance, event or occurrence as it actually takes or took place; a physical object or appearance, as it usually exists or existed. An actual and absolute reality as distinguished from mere supposition or opinion. A truth, as distinguished from fiction or error. "Fact" means reality of events or things the actual occurrence or existence of which is to be determined by evidence.

Evidence is defined as:

> All the means by which any alleged matter of fact, the truth of which is submitted to investigation, is established or disproved. Any matter of fact, the effect, tendency, or design of which is to produce in the mind a persuasion of the existence or nonexistence of some matter of fact. That which demonstrates, makes clear, or ascertains the truth of the very fact or point in issue, either on one side or on the other. That

> which tends to produce conviction in the mind as to existence of a fact. The means sanctioned by law of ascertaining in a judicial proceeding the truth respecting a question of fact.
> The rules of evidence control the presentation of facts before the Court. Their purpose is to facilitate the introduction of all logically relevant facts "without sacrificing any fundamental policy of the law which may be of more importance than the ascertainment of the truth" (Sopinka et al. 1992).

The law of evidence regulates:

1. what matters are admissible before the Court; and
2. the method by which admissible facts are placed before it.

The role of the consulting arborist is to provide the client or the client's lawyer and the judge or judge and jury hearing the case with evidence that will assist in determining what is factual. The arborist states what was found at the site, along with his or her opinion about an array of factors pertinent to the case. The judge then decides which parts of the evidence submitted are to be considered the facts of the case, and how these facts relate to the incident in question. Emanating from the facts and their relevance comes the judge's interpretation of how this information bears on the case and who is liable, and to what extent. In all cases, the judge or judge and jury is the "trier of fact." The arborist is there to provide an opinion as to what these facts should be; that is, to assist the trier of fact in making a determination.

It is not unusual for opposing sides to have arborists claiming widely different sets of facts. Such divergence in expert opinion exists for many reasons, such as different: levels of experience, interpretations, assessment methodologies, and degrees of subjectivity. It is up to the lawyers to test the experts to see which can most credibly defend his or her opinion.

Arborists gather and assess some straightforward evidence, such as damage resulting from blatant trespass. Other types of evidence may be much more difficult to ascertain, such as: acceptable loss of value when trees are removed, exact causes of tree failure, or whether certain characteristics of the tree would have been apparent prior to failure. In all cases, however, there are certain procedures that must be followed by the arborist if the evidence and report are to stand up to scrutiny in a legal arena.

10.3 Goals of the Canadian Trial System

The search for truth, enhancement of the efficiency of the process, and concern for fairness to the parties are all goals which centre on the trial process itself. In particular, to search for truth is the essential purpose of the trial system in Canadian society. This goal can be elusive because the system is an adversarial one where fact presentation is controlled by litigants and their counsel.

10.4 Rules of the Law of Evidence

Apart from the deficiencies in evidence caused by the adversarial framework, the evidence itself may be inherently weak due, for example, to witnesses' imperfect recollection or perception. In an attempt to limit these frailties, many of the rules of evidence are concerned with ensuring the reliability or accuracy of the evidence the Court receives. An example is the hearsay rule, which seeks to exclude evidence from the Court when the evidence is derived from statements made by a person who is not a witness. This rule is based in part on a fear of fabrication and on a reluctance to receive evidence where the perception of the actual witness of the events cannot be tested in cross-examination.

The rules of evidence are made up of common law principles, statutory provisions, and constitutional principles. The common law of evidence is derived from the decisions of judges in England and Canada, and in some cases, the United States and other common law countries. Statutory rules are included in statutes such as the *Canada Evidence Act* for *Criminal Code* and federal court matters, and in the various provincial *Evidence Acts* for most other civil litigation. These statutes amplify the common law rules. In the context of expert evidence, the *Evidence Acts*, federally and provincially, provide directions on the timing and format of expert reports.

In addition, practice rules issued by provincial courts may also have an impact on the rules of evidence. For example, in British Columbia, Rule 40A of the *Supreme Court Rules* contains detailed directions on the timing and format of expert reports. Lawyers must comply with these rules or the courts may rule that the reports are inadmissible.

Evidentiary rules also address interests outside of the trial process. For example, there is a large body of law, both common and statutory, that governs the production of documents belonging to or in possession of the government. Exclusion of such materials could sometimes be justified by a need to protect national security or the processes of government.[4]

10.5 The Preliminaries

10.5.1 Understanding the Client's Needs

Arborists should never reveal details of their knowledge of or opinions about a case until they have a letter or contract stating that they are retained, with conditions of work and remuneration clearly noted. Arborists may on occasion be approached by lawyers who will ask in general terms for information on a case or suggestions for approaches to a problem. Arborists should avoid commenting on these questions for two reasons: 1) they may be quoted later on this or another issue and find that what they initially said was inaccurate due to the absence of full details; or 2) they may find that, because they have discussed the case, albeit in only very general terms, they can't work for the other side because their information may be seen to be tainted or biased in some manner. There is also the fundamental issue of a fee for services rendered.

Before consulting arborists even commence work, they should understand what the client wants, why, when, and for what purpose. Arborists should not hesitate to ask questions that ensure a thorough understanding of the task before starting work. If the case is being discussed with a lawyer representing a client, arborists should be sure to ask questions about the exact nature of the case.

It is important to be sure the client clearly understands:

- how much will be charged and on what basis (hourly, daily, disbursements etc);
- what work will be undertaken;
- when the work will be performed; and
- what deliverable products can be expected

A written contract is not always simple to achieve on small projects where a homeowner requires immediate advice. At the very least, arborists should ensure there has been clear verbal communication on the above points, and then refer back to them in their report or letter. However, a formal letter of agreement is better and more prudent. If such a letter is used, make sure that the client acknowledges in writing that the proposal is accepted. Be sure that permission is given to visit the site in question. If the site involves two properties, be sure that permission is obtained from both property owners. If the task involves hazard tree inspections, clearly explain in writing to the client what will be done. If the inspection will not include a root collar excavation, say so. If further inspection work is required, such as climbing up into the crown or excavation of the root collar, say so.

Arborists who provide specifications for someone else's use should be sure to note that they accept no liability or responsibility for the manner in which the specifications are implemented, unless they can specifically control, supervise, and inspect the implementation with authority (that is, the arborist can direct the actions to be taken, changed, or even stopped). It should also be noted that there is a difference between the terms review and inspect. Consultants may review the work as it proceeds, but inspection is work reserved for authorised agencies or personnel; that is, people with the authority to accept or reject the work as it proceeds or when it is completed. Arborists' contracts should specify which role they are assuming. In the event they are undertaking inspection work, they need to be sure they have authority to ensure the quality of work.

Arborists' contracts should specify whether they are reviewing or inspecting the work

While there are no known cases of arborists being sued over lack of supervision, the engineering world provides an indication of what can happen. In the case of *Homes by Jayman Ltd. v. Kellam Berg Engineering & Surveys Ltd.* (1995), the engineering company that designed a septic system (Kellam Berg), and the contractor who built it were found jointly liable when the system failed. The engineer had no written contract to supervise or inspect the construction work, the contractor installed it incorrectly, and the plaintiff subsequently damaged it. The judge found that although there had been no specific agreement about supervision, it was standard practice in other similar endeavours for the engineer to be involved in supervisory roles. Thus, the judge found that the developer could have reasonably expected the engineer to inspect the project as it was built, which implied a certain degree of supervision. Hence, the engineer did owe a duty of care to supervise the work, and therefore had to share in the payment of damages.

10.5.2 Working with the Lawyer

The lawyer retaining an arborist as an expert should ensure that the arborist is given sufficient and correct terms of reference. There are two important reasons for ensuring the sufficiency and correctness of the terms of reference. First, the lawyer controls the arborist's expenditures by precisely defining the issues to be investigated. Secondly, the arborist needs sufficient instructions to be able to carry out a proper and adequate assessment.

The lawyer's letter defining the terms of reference should be brief and advise that the arborist's assessment or opinion should be objective. An example of the type of instructions that should be contained in such a letter is provided below.

> This letter will confirm the writer's instructions to you on behalf of our client, Johnson Enterprises Ltd., with respect to your preparation in writing of an objective assessment of the falling of a large coniferous tree on the residence located at 1250 Burnaby Way, Burnaby, British Columbia. This assessment should be forwarded to the attention of the writer by June 19, 1984.
>
> We have advised Mr. Johnson, the President of Johnson Enterprises Ltd., the building owner, that you may wish to review this matter with him and with some of his employees. Would you kindly make your own arrangements with Mr. Johnson.

> We have assembled the documents we have obtained from Mr. Johnson, the Defendant and third parties. These documents are available for your inspection at our offices at your convenience.
>
> It is our understanding that your fee for services rendered with respect to the preparation of the report and attendance at the trial will be a fee of $XX.00 per hour plus all reasonable expenses. It is estimated that the budget for all services, including the preparation of your report and one day of testimony at trial, amounts to $X,000. It is agreed that your fee will be paid by Johnson Enterprises Ltd., and that you will submit your statement of account to Mr. Johnson on a monthly basis.

For the lawyer, it is important to confirm the manner in which the arborist's fees will be paid. If the fees will be paid directly by the client and not by the lawyer, the arborist should make sure the monies owing can be collected. The best arrangement seems to be a written agreement that the expert is paid by the lawyer on behalf of the client. In this way, there is always some legal recourse in the event of non-payment. The expert report (and all related communications) should always be directed to the attention of the lawyer so that it becomes a privileged communication (under the umbrella of solicitor-client privilege) and can be protected to some degree from forced disclosure on the witness stand.

If the case involves testifying, the arborist should avoid contingency-based fees that add a certain percentage in the event the case is won

If the case involves testifying, the arborist should avoid contingency-based fees that add a certain percentage in the event the case is won.

10.5.3 Budgeting

Another important matter for the lawyer to consider is the cost of retaining the arborist's services. Experts are normally a necessary but often a very expensive component of any complex litigation case. It is therefore important to control the client's expenditures by establishing a budget for the expert.

The correct procedure for the lawyer is to provide the arborist with preliminary details and documentation. The lawyer should request that the arborist review the issues and probable terms of reference and then provide an estimated budget. The lawyer can then review the budget with the client and obtain approval.

The lawyer should provide a schedule of critical dates, including a deadline for the report and the anticipated trial date.

If the lawyer decides to retain the arborist for a major report, he or she should discuss the overall cost of services, including:

1. Cost of the preliminary report;
2. Cost of the final report;
3. Cost of reviewing opposing expert's report;
4. Preparation for trial; and
5. Attendance at trial.

10.6 On the Site

10.6.1 Upon Arrival

Upon arrival at the site, arborists should acquaint themselves with the person in charge by presenting their business card (letter of authority if needed) and reviewing why they are there and what they intend to do. In many cases, an arborist's arrival on site or the arrival of documents for review will bring up new issues not previously discussed. At all times, keep the client fully informed of any changes or detours that may be necessary, and make sure that each side acknowledges them in writing (or get a verbal acceptance and follow up with a written confirmation). Arborists should not give hard opinions on site. If asked, arborists should be sure to qualify their opinions with caveats that allow them to think, analyse, and carefully respond in writing later on. Take down the facts, take any measurements needed, record all the details, and then allow some time for reflection and report writing. Giving a preliminary opinion can lead to problems if, on more detailed analysis, the opinion changes, especially if anyone takes the preliminary opinion as final and initiates other actions on that basis.

Giving a preliminary opinion can lead to problems if, on more detailed analysis, the opinion changes...

While working on site, arborists may well encounter a dangerous (hazardous) tree that was not part of their original mandate. Without exception, the client or owner should be notified in writing (verbally at the time if possible); otherwise, arborists expose themselves to liability. As professionals, arborists are able to identify problems that the layperson might not see. Arborists who do not report such problems may be charged with professional negligence because either they should have seen the problem or they should have reported it or both. Arborists should always be explicit about what they see and put it in writing. The first site visit should be structured as though it will be the only one—don't rely on a second opportunity.

10.6.2 Keeping Good Records

Arborists should keep notes of each site visited. These field notes are the backbone of any subsequent work and should not be rewritten later on to clarify things. Adding other notes after the field visit to clarify the original ones is fine, but it may be wise to use a different colour of ink to clearly separate the original from the additions. In several instances, consultants' notes from the field have been subpoenaed in court, only to be discarded because they were rewritten. Even if no changes were made, the possibility of tampered evidence does nothing to strengthen the arborist's credibility. Neatly written notes on spotless paper are not normal for most field conditions; somewhat scruffy notes, that show signs of field conditions (rain drops, dirt, resin) are more typical.

Often, there is only one opportunity to visit a site, so in advance prepare a list of things to do and inspect.

Keep notes about times on and off site, weather conditions, people seen and/or talked to, and any other incidental items of relevance. For example, note if any other work is taking place in the area that might affect the evaluation, and note all the work that has apparently taken place already (particularly important on construction sites). As another example, if the arborist (or the contractor being supervised) drops branches into a fish bearing creek and then extracts them, there is a possibility of an infraction. If the arborist's notes reveal that the creek was running high and was very heavily laden with silt at the time, it can then be argued that the infraction was trivial relative to the condition of the stream at that time.

If arborists neglect to make such notes (and better still, take photographs), then they have no solid evidence to back up what they saw. Under cross-examination, it may then be possible for the opposing lawyer to make it look as if a less than thorough inspection was carried out. The judge (or judge and jury) only has the evidence presented to them. They do not have an arborist's intuition and experience of the site conditions, and may not understand the implications of technical terms. Be observant and record the usual and the unusual.

10.6.3 Taking Photographs

Photographs are useful. It helps to develop a clear sense of what needs to be photographed before arriving on site. Once there, arborists should take lots of photographs from different angles, some close up, and others more distant. Document each photograph and ensure that the directions of view and distances are recorded (for example, looking east 100 metres away from property line). If the photographs are not the arborist's own, their origin should be noted, together with any assumptions made about them. Always work with an original photograph. Digital images are very easily manipulated, and cannot be considered a reliable source of evidence for analysis unless there is an authenticated chain of handling from the person who took the picture through to the time it was delivered to the arborist. Even so, the arborist should note that the image examined was in digital form and the original could not be verified.

10.6.4 Maintaining Objectivity

In many cases, clients believe that they are absolutely correct. High levels of anger and indignation are common among clients who believe they have been wronged in some manner. This emotionally charged atmosphere often makes it difficult for the consultant to concentrate on the work in hand, and it is easy to become distracted or worse still, biased, by the opinions and attitudes of the client. A good example is the issue of boundary lines and trespass. Just because the client is adamant that the line is here and the trees cut were his does not mean that he is right. Is there a legal survey pin that can be worked from? Has a recent survey been undertaken? Is the arborist trespassing while working? If there is any possibility of trespass, the arborist should be sure to obtain written permission to go on the other person's property before arriving at the site. If there are doubts, make notes about them and make sure the client understands why there are misgivings. It is better for all concerned to find out mistakes before going into court.

Unless the arborist is a registered surveyor, he or she should not attempt to undertake detailed survey work that will then be the foundation for further work. Let the professional surveyor demarcate the line, then undertake an accurate assessment. If the client does not want to undertake the expense of a survey, be sure to note in the report that the work done was based on the client's determination of property lines, and that there are no guarantees as to the accuracy of the client's knowledge. Similarly, when a client points out certain features and states categorically that they are theirs or were built at such and such a time, arborists should note the source of these facts as belonging to the client and not verified by them.

10.7 The Report

The style of report required by a lawyer may vary from company to company. There are, however, certain fundamental aspects that must be incorporated into each and every report.

10.7.1 Just the Facts

The written report must contain factual information only. There is no room in the report for emotion and personal biases. In the face of uncertainties, either state the limitations of the conclusion and why they exist, or leave the speculation out all together. Differentiate between factual evidence (the tree was x centimetres in diameter and y metres long) and opinion (in my opinion it fell down due to damaged roots on the north side). Write a clear and unambiguous report, state the facts clearly and provide all the supporting evidence. Any calculations undertaken must be explainable and defensible. Show the figures used and how they were incorporated in an appendix. Check to make sure the figures and calculations are correct.

The basic contents of any report should include the "seven basic elements" listed by Abeyta (1995). These are:

1. Define the assignment
   - What is it?
   - What limits are there?
   - The background or history of the assignment
   - The purpose and use of the report
2. Note the observations made.
   - Describe the on site investigations
   - Describe any historical or literature research used
   - Note which parts are based on your expertise and common knowledge
3. Describe any analysis or testing undertaken
4. Discussion
5. Conclusion
6. Recommendation(s)
7. Supporting Information
   - Describe qualifications and experience relevant to the assignment

Although an effective report may not have all of the above headings, each of those elements should be contained in some part of the report. It is also extremely important to have, at the beginning of the report, a summary of the arborist's assessments or opinions, including the purpose of the report and brief conclusions. The purpose of the summary at the beginning is to give the lawyer the opportunity to advance the arborist's opinion when the Court is most interested—at first introduction.

10.7.2 Writing Style

Be careful with the use of language. For example, a statement such as, "a detailed examination was made….," must be defensible. What was done must in fact have been detailed and not just routine. Questions under cross examination might include, "What is detailed ? How detailed is detailed? How much more could you have done?" If it seems that a lot more could have been done the judge may be left with the opinion that what was done was not detailed at all, and the case may appear to have been overstated. To make an effective and persuasive case, factual evidence is needed. For example, if in the arborist's opinion structural damage has been caused by the neighbour's tree roots, then the roots damaging the structure must be convincingly proven to be the same roots attached to the neighbour's tree. Anything less will be conjecture and may be discarded by the courts (see, for example, *Canadian Red Cross Society v. Past* [1979]).

The report should be completely objective and impartial. The arborist should attempt to remove all adjectives and adverbs from the report unless absolutely necessary.

Once the draft report is written, the lawyer should review statements, reports, documents and articles which the arborist previously wrote to determine if the opinion being given now conflicts with previous positions or opinions. Note that arborists also have a responsibility to be aware of what they have previously written or said, including a record of the expert's deposition.

The arborist's prior experience and expertise should be developed to ensure that the qualifications are fully explained and emphasised. Make sure the resume or statement of qualifications is current and correct.

10.7.3 Confidentiality

It is safe to assume that there is never such a thing as a confidential report. The rules of evidence vary, but arborists should always assume that any written document may be seen by other people. The guiding maxim in all report preparation should be: be prepared to defend your report in court. If you cannot defend it, don't write it.

In British Columbia, the Court has held that once a party calls an expert witness to testify, either in person or through the filing of a report under section 10 of the B.C. *Evidence Act* (or Rule 40A of the B.C. *Supreme Court Rules*), all solicitor-client privilege which may earlier have attached to documents prepared for litigation is lost (*Vancouver Community College v. Phillips Barrett* [1987]). Similar provisions may apply in other provinces and should be checked before assuming all aspects of solicitor-client privilege are in force.

Arborists should be aware that when they are called to testify, the opposing counsel may request copies of all documents used in preparation of their report, including all drafts of the report. In cross-examination, arborists may be asked to explain the reasons for any changes in opinion observed between the draft and final reports.

In light of the *Phillips Barrett* case, what should be done with draft reports, papers, and notes? One position is that all draft reports, documents, papers, notes or anything else reflecting changes in the arborist's opinion should be preserved. Another position is that all drafts should be discarded as soon as they have been revised so that only the arborist's field notes and final report remain.

The lawyer with whom the arborist is working may respond to the first report with suggestions for changes, perhaps a re-ordering of points, or some additions and deletions. The manner in which the arborist responds to these suggestions is important. Above all, the arborist has to be able to defend the final report under cross examination. The arborist should not write something to please a lawyer if it cannot be defended.

10.7.4 Using Waivers

A very important part of any report, especially when the findings have the potential to be used out of context, is a waiver that states the report can only be used by the original client for a specific case, with written permission to be obtained from the author before any other use and distribution of the report. A recent and important court case in Canada is *Wolverine Tube v. Noranda* (1995), in which the environmental consulting company had in its reports a waiver stating that the reports were not to be relied upon by a third party. The vendor of a site provided the purchaser with copies of the reports. The purchaser subsequently found other issues on site which they disliked and sued the vendor. The consultant was dragged in but exonerated because of the waiver, which preceded the text in all three volumes and was recognized, understood, and ignored by the client. Had the waiver not been present, the consultant may well have been exposed to considerable liability.

Similarly, hazard tree assessments, regardless of whether they are to be used in court (there is always that possibility), should always include a waiver about the limitations of the assessment. An example of a hazard tree assessment waiver is provided below.

> **Limitations of this Assessment**
>
> It is our Company's policy to attach the following clause regarding limitations. We do this to ensure that developers or owners are clearly aware of what is technically and professionally realistic in retaining trees.
>
> The assessment of the trees presented in this report has been made using accepted arboricultural techniques. These include a visual examination of the above-ground parts of each tree for structural defects, scars, external indications of decay such as fungal fruiting bodies, evidence of insect attack, discoloured foliage, the condition of any visible root structures, the degree and direction of lean (if any), the general condition of the tree(s) and the surrounding site, and the proximity of property and people. Except where specifically noted in the report, none of the trees examined were dissected, cored, probed, or climbed, and detailed root crown examinations involving excavation were not undertaken.
>
> Notwithstanding the recommendations and conclusions made in this report, it must be

> realised that trees are living organisms, and their health and vigour constantly change over time. They are not immune to changes in site conditions, or seasonal variations in the weather conditions.
>
> While reasonable efforts have been made to ensure that the trees recommended for retention are healthy, no guarantees are offered, or implied, that these trees, or any parts of them, will remain standing. It is both professionally and practically impossible to predict with absolute certainty the behaviour of any single tree or group of trees or their component parts in all circumstances. Inevitably, a standing tree will always pose some risk. Most trees have the potential for failure in the event of adverse weather conditions, and this risk can only be eliminated if the tree is removed.
>
> Although every effort has been made to ensure that this assessment is reasonably accurate, the trees should be re-assessed periodically. The assessment presented in this report is valid at the time of the inspection.

The lesson here is to be very explicit regarding what can be done with the arborist's work, and what limitations are placed on it.

10.7.5 Storing Records

All records, field notes, and photographs (and negatives) that have been used to prepare the final report should be retained in an orderly manner for at least five years after the assignment is completed (ASCA 1996). A court case typically is processed one to two years after a claim is filed. If the case is then appealed, another few years may pass. There is always a chance that an arborist who was involved may later be asked for information.

10.8 Knowing the Limits

As with all areas of specialization, there are limits to what the arborist knows. The arborist can respond in three ways to a lack of knowledge. One is to research the issue and derive what is believed to be the answer. A second choice is to bring in another person who has more experience in the area in question. A third and more dubious choice is to try to produce an answer without fully understanding the problem. Hendricksen (1996) categorised these approaches in terms of competence:

1. *Unconscious Incompetence—"I don't know that I don't know."*
   In other words the arborist has not even recognized that there is a problem and is proceeding under the potentially false assumption that all is well.
2. *Conscious Incompetence—"I know that I don't know."*
   Now the arborist has recognized his or her limits and can decide what to do.
3. *Conscious Competence—"I know but I have to think about it."*
   The arborist has the knowledge required but needs to brush it up a bit before it can be used effectively.
4. *Unconscious Competence—"I know and am so good at it that I don't even have to think about it."*
   This may be true, and is the hallmark of someone who is very good at their chosen field of endeavour, but, it is also potentially dangerous, since familiarity can breed contempt.

The best level of competence lies somewhere in between option three and option four. Arborists should recognize their limits of competency and strive to exceed them so that new knowledge becomes second nature, but not get so confident that they overlook some vital aspect of the problem. Overconfidence can be just as risky as ignorance.

10.9 Expert or Opinion Evidence

Generally speaking, lay witnesses at a trial are not permitted to give their opinion. A lay witness normally may speak only on factual matters which he or she has personally observed. In order to provide the Judge or jury with the necessary technical or scientific basis upon which to properly assess the evidence presented, the Courts recognize an exception to the opinion rule for expert witnesses.

Originally, expert evidence was permitted only in cases where the subject matter in question was beyond the capabilities of experienced persons, so that they could not form a correct judgement without such assistance.

Dixon, J., in *R. v. Abbey* (1982), described the role of the expert in the following way:

> With respect to matters calling for special knowledge, an expert in the field may draw inferences and state his opinion. An expert's function is precisely this: to provide the Judge and jury with a ready-made inference which the Judge and jury, due to the technical nature of the facts, are unable to formulate. "An expert's opinion" is admissible to furnish the Court with scientific information which is likely to be outside the experience and knowledge of a

> Judge or jury. If on the proven facts a Judge or jury can form their own conclusions without help, then the opinion of the expert is unnecessary. [*R. v. Turner* (1974), 60 Crim. App. R. 80 at page 83, per Lawton, L.J.]

The admissibility of expert evidence does not depend upon the means by which the skill is acquired. As long as the Court is satisfied that the witness is sufficiently experienced in the subject matter at issue, the Court will not be concerned about whether his or her skill was derived from specific studies or practical training, although its origin may affect the weight given to the evidence.[5]

10.10 Desirable Characteristics of an Expert Witness

The ideal expert should have many qualities, including writing skills, knowledge of the issues, and accessibility to counsel. In the view of one leading British Columbian trial lawyer, the most important quality for an expert to have is to be a "persuader."[6] That is, the expert should be able to convince the other side, the other expert, and in particular, the Judge, of the veracity of his or her opinion.

Some other desirable characteristics of an expert witness are:

1. a thorough knowledge of the scientific principles involved;
2. considerable experience in the application of these principles;
3. the ability to present facts and opinions in a clear, concise, and convincing manner and in terms readily understood by non-technical persons;

4. appreciation of both sides of a question and the ability to function in an impartial, unbiased manner, and in fairness to all concerned;
5. reasonable knowledge of and preferably some previous experience in Court procedures; and
6. willingness to have one's opinion or judgement questioned by all counsel.

10.11 The Expert at the Trial

It is extremely important for the lawyer to prepare the expert witness for trial by advising him or her of the procedures that will be followed. This preparation includes a "dress rehearsal" of anticipated questions in examination and cross-examination.

The following are some instructions that the lawyer should give the expert witness in preparation for his or her attendance at the trial:

1. tell the truth;
2. do not anticipate the question;
3. ensure you understand the question;
4. speak slowly and clearly to facilitate court reporting, and to allow time for the judge to take notes and absorb your important testimony;
5. do not reconstruct the facts;
6. do not volunteer information;
7. do not agree with suggestions posed by opposing counsel unless the question clearly and accurately reflects the facts and your opinions;
8. do not argue with counsel;

9. remember that even the best cross-examiner can do less damage to a witness who is telling the truth;
10. if you need to take notes and reports into the witness box, make sure that they are well organised and that you are willing to share them with the court if requested;
11. do not make inferences or give opinions on areas that are outside your field of expertise; and
12. dress conservatively and avoid flashiness so that the judge can concentrate on your testimony, not your attire.

The expert witness should be cautioned against excessive frankness or false modesty. This comment relates to the problem that arises when a witness, on being asked whether he or she is in fact an expert, testifies that he or she may not be one when compared to another expert. This type of answer usually occurs when the expert is confronted with the experience of perhaps another more qualified person giving evidence. It is important for the lawyer to advise the expert witness in advance of the trial that he or she is indeed knowledgeable and experienced. If questioned in this matter, the expert witness should confirm his or her expertise for the Court.

The simplest method for the lawyer to present the expert witness to Court is to call the expert as witness, have him or her sworn and identify him or her. Expert witnesses are usually qualified before the court, which involves questioning them about their background and qualifications. At the completion of this evidence, the lawyer should ask the Court to qualify the witness as an expert. In the case of an expert witness with more than one expertise—for instance, as an arborist, a planner, and a registered professional forester—it may be necessary to define specifically the area of expertise relevant to the trial, such as an arborist having expertise relating to hazardous trees or a planner with specific expertise in heritage trees in the context of subdivision planning.

The Court will normally ask opposing counsel if they have any questions about the witness' qualifications. Opposing counsel may choose to cross-examine the expert witness on his or her qualifications and object to him or her being qualified as an expert.

Once the expert witness has been qualified before the court, his or her expert report and any associated materials will be produced and marked as exhibits to facilitate reference and understanding during examination in chief (when the lawyer with whom the witness has been working asks questions) and cross-examination (when the opposing lawyer asks questions). During examination in chief, the lawyer will likely refer to the witness' report, read it to the Court, and ask the witness to elaborate, if required.

10.12 Preparing an Expert's Report

Generally speaking, the lawyer should not commence examination for discovery of the opposition's witnesses prior to completion of the expert witness' draft report. In most expert reports, the expert's opinion will be a response to facts of which the expert does not normally have first-hand knowledge or experience. In most instances, the lawyer will have to ensure that the facts and assumptions contained in the expert's report are proved by documentation or admissions obtained on examination for discovery. The relevant documents should then be referenced in the final draft of the expert's report. If the documents are few in number, they should be included as an appendix to the expert's report; if voluminous, they should be compiled in a separate assembly, tabulated, and identified in the narrative of the expert's report as reference documents upon which the expert relies for the factual foundation.

In accordance with B.C. *Supreme Court Rules*, Rule 40A, it is recommended that the expert's report contain at least the four parts discussed below,[7]

10.12.1 Stating Assumed Facts

The lawyer will prepare and deliver to the expert witness a Statement of Assumed Facts together with any relevant documents. These materials set out what facts the expert witness knows and how they are known. For example, if the expert's evidence refers to a site plan, the origin of the plan must be stated as shown below.

1.1 I reviewed a lot plan of the area upon which the subject property is indicated.

1.2 The lot plan referred to above was prepared by ___________ in 19_ on behalf of the City of _________, and is assumed to be authentic and accurate.

In forming an opinion, the expert witness should not draw inferences from any documents or facts. The expert witness should base his or her opinion on the assumptions that the lawyer has provided and on nothing else. It may be inappropriate for a lawyer to provide the expert witness with access to discovery transcripts, witness statements, or other evidence if the information might bias the expert's opinion. The lawyer should state that the facts will be established at trial. Other rules that the lawyer should follow in order to comply with legal requirements and assist the expert witness are:

1. The Statement of Assumed Facts will be factual and not argumentative
2. Any facts which bear on the expert's opinion, both for and against the client's position, will be set out as assumptions. The lawyer may set out alternate sets of facts to guard against the possibility that the Court's findings may not be consistent with the hypothetical facts
3. The lawyer may not cite the source of the assumed facts
4. The expert may request any additional necessary facts
5. The lawyer will present the Statement of Assumed Facts in brief numbered paragraphs
6. The importance or unimportance of the assumed facts to the expert's opinion may be indicated by the expert in his or her report
7. The expert may sometimes review documents that he or she considers important or relevant, but this matter the lawyer must

consider on a document by document basis since it is important that the expert does not do so in order to make findings of fact.

The lawyer may ask the expert witness to point out inferences that could be used in cross-examining the other side's witnesses, but such activity must be kept completely separate from preparation of the expert report.

10.12.2 Questions For The Expert

It is advantageous for the lawyer to draft and submit to the expert witness the written questions to which he or she is to respond in Court. These questions will be fine tuned in consultation with the expert. The expert should quote the final draft in his or her report.

10.12.3 The Expert's Opinion

The third part of the report will be the expert's opinion. The lawyer should be very careful not to interfere in any way with the substance of the expert's opinion—he or she must not influence the expert's objectivity.

The expert's opinion has certain parameters. For expert testimony to be admissible, it must be necessary to assist the trier of fact to form a judgement. The expert cannot assume the role of the Court or a jury by expressing an opinion on the issue to be decided (known as the ultimate issue). The expert must not:

1. evaluate conflicting evidence;
2. make findings of fact;
3. allocate fault or responsibility;
4. interpret contracts or other legal documents;
5. advance factual or legal arguments; or
6. state legal conclusions.

The expert's opinion must be scientific or technical in nature. The expert should not discuss the legal or general merits of the case. In some cases, the Court has disallowed expert evidence as an attempt by counsel to advance argument in the form of expert evidence.

10.12.4 The Expert's Qualifications

In addition to attaching a copy of the expert's resume to the report, the lawyer should request that the expert prepare a summary of relevant experience and qualifications for the purpose of this report. The expert should elaborate on the nature and extent of his or her particular experience in the area in question.

10.13 Other Matters

10.13.1 Professional Liability Insurance

In Canada, certified arborists do not have the option of using the American insurance coverage available to arborists practising in the United States. If arborists are going to undertake court cases, hazard tree inspections, and professional reports that contain a clearly stated opinion, they should have professional liability coverage (also known as errors and omissions coverage). Beware of coverage offered by other professional associations, such as professional foresters or professional agrologists. It may not be sufficient. For example, one of the authors of this book is both a professional forester and a certified arborist. Coverage under the umbrella of professional forestry was available at half the price of a broader option available to environmental consultants. However, upon detailed enquiry it became clear that the forester's coverage would only apply if the consultant had been specifically hired as a forester and not as an arborist. Moreover, many of the tasks performed by foresters were seen differently from those performed by arborists, and were not present on the list of accepted practices held by the insurers for the foresters. Hence, the lesser price of the annual premiums could in fact lead to absence of coverage in the event of a claim. An absence of coverage always has the potential to bankrupt the arborist's business. If the business is not an incorporated or limited company, the arborist might also lose personal assets (house, car, other possessions) if the court presses for compensation to meet a claim. When arborists take out their professional insurance, they should write out all the tasks they do or expect to do and be sure to get in writing that this list of activities is covered. Be sure that any exclusions are understood.

10.13.2 Signatures and the Board of Directors

Under Canadian law both the employee signing a report and the Board of Directors guiding the company may be liable for the work undertaken. While no cases were encountered that involve arboriculture, there have been some very high profile environmental contamination cases wherein the Company directors were found liable for site problems on the basis that they had a responsibility, and a duty of care, to be fully cognizant of their company's operations and standards. When problems arose, they did not respond properly due to lack of understanding. In one case, the Ontario Provincial Court prohibited the company from paying the directors' fines, which then became the director's personal liability (*R. v. Bata Industries Ltd.* [1992]; subsequently overturned in the court of appeal in 1995). In some environmental prosecutions, directors have been jailed as well as fined.

It is vital to ensure that employees sign reports properly. The sign-off should state, right under the "Yours truly" line (or whatever sign off format you use): On Behalf of <the legally registered company name in full as it appears on the registration form>. This line is then followed by the signature, and under that, the printed name of the person signing. This format ensures that the work is clearly the property of the company, not the individual.

If the arborist operates under a non-incorporated business or sole proprietorship, he or she should be extremely careful about the type of work undertaken and the statements made because he or she is highly vulnerable to personal liability. It is common for everyone associated with a project to be added as defendants in major lawsuits. In effect, the writ names everyone and then lets each named person or business demonstrate why they should not be involved. Proving non-involvement can be a time consuming, stressful, and expensive undertaking.

10.14 Summary

Given the increasing number of court cases involving trees, consulting arborists will have a continuing presence in the courts. The consulting arborist can fulfill an important function in assisting the courts to decide what evidence will be accepted in a tree related case. There are two roles that the consultant can adopt. One is as a consulting expert who may act as an advocate. The other is as an impartial expert witness who states the facts and renders an opinion to assist the judge or judge and jury in their capacity as the trier of fact.

In both roles it is vitally important to maintain a high standard of professionalism. Though these standards are changing all the time, the recently adopted **Standards of Professional Practice** being promulgated by the American Society of Consulting Arborists provides a good guide.

Appendix
Table of Cases

Case	Section
Amos v. New Brunswick (Electric Power Commission), [1977] 1 S.C.R. 500, 8 N.R. 537, 13 N.B.R. (2d) 207, 70 D.L.R. (3d) 741; reversing (1975) 10 N.B.R. (2d) 644, 60 D.L.R. (3d) 495 (C.A.); reversing (1974) 9 N.B.R. (2d) 358 (Q.B.).	6.3
Anderson v. Skender (1991), 61 B.C.L.R. (2d) 292, [1992] 1 W.W.R. 709 (S.C.); varied (1993), 84 B.C.L.R. (2d) 135, (1993), 17 C.C.L.T. (2d) 160, 32 B.C.A.C. 142, 53 W.A.C 142, [1994] 1 W.W.R. 186 (C.A.); varied on reconsideration *Anderson v. Skender (No. 2)*(1993), 36 B.C.A.C. 79, 58 W.A.C. 79 (C.A.); leave to appeal to S.C.C. refused [1994] 2 W.W.R lxiv (note), 18 C.C.L.T. (2d) 324 (note), 87 B.C.L.R. (2d) xxxii (note), 166 N.R. 238 (note), 47 B.C.A.C. 255 (note), 76 W.A.C. 255 (note) (S.C.C.).	2.2.1 4.2 4.3.3
Anns v. Merton London Borough Council, [1978] A.C. 728, [1977] 2 W.L.R. 1024, [1977] 2 All E.R. 492 (H.L.).	1.2.3
Austin v. Rescon Construction (1984) Ltd. (1989), 48 C.C.L.T. 64, 36 B.C.L.R. (2d) 21, 57 D.L.R. (4th) 591, [1989] B.C.W.L.D. 1334 (C.A.), reversing (1987) 18 B.C.L.R. (2d) 328, 45 D.L.R. (4th) 559, 45 C.C.L.T. 137, [1987] B.C.W.L.D. 3618 (S.C.).	2.2.2
Barnstead v. Ramsay, [1996] B.C.W.L.D. 1528; additional reasons [1996] B.C.J. No. 970, [1996] B.C.W.L.D. 2489 (S.C.).	1.2.3; 3.4.1 3.6
Bawa v. Noton, [1996] B.C.J. No. 1703, [1994] B.C.W.L.D. 1941; appeal allowed in part [1996] B.C.W.L.D. 1025, (1996), 74 B.C.A.C. 154, (1996), 121 W.A.C. 154 (C.A.).	2.3.2

Case	Section
Black v. Zager (1982), 18 Man. R. (2d) 22, 22 C.C.L.T. 231 (Q.B.).	4.3.3
Bosch v. Hudson, unreported, Vancouver Registry No. C960359 (B.C.S.C.).	8.2
Bottan v. Richmond Hill (Town), [1995] O.J. No. 3501 (Ont. Gen. Div.).	5.6
Bottoni v. Henderson (1978), 21 O.R. (2d) 369, 90 D.L.R. (3d) 301 (H.C.).	8.4; 8.5
Boudreau v. T.P. Downey & Sons Ltd. (1988), 91 N.B.R. (2d) 330, (1988) 232 A.P.R. 330 (C.A.); reversing [1987] N.B.J. No. 858 (Q.B.).	2.2.1 3.4.1
British Columbia Telephone Company v. Cooper Builders Ltd., unreported, March 11, 1993, Coquitlam No. C034453 (B.C. Prov. Ct.).	6.2
Buckland v. Guildford Gas Light & Coke Co., [1949] 1 K.B. 410.	6.3
Buie v. Etobicoke (City) (1984), 25 M.P.L.R. 70, 31 R.P.R. 107 (Ont. Prov. Ct.).	4.3.1
Butler v. Standard Telephone and Cables Ltd., [1940] 1 K.B. 399, [1940] 1 All E.R. 121.	4.3.1
Caminer and Another v. Northern and London Investment Trust Ltd., [1950] 2 All E.R. 486 (H.L.).	8.4
Campbell v. Calgary (City) (1984), 55 A.R. 73 (Q.B.).	5.2
Canadian Red Cross Society v. Past, unreported, November 22, 1979, Victoria Registry No. 001859/78 (B.C.S.C.).	4.3.1
Carr-Harris v. Schacter, [1956] O.R. 994, (1956) 41 D.L.R.(2d) 225 (H.C.).	2.3.2; 3.3
Centrum Land Corp. v. Institute of Chartered Accountants of Ontario (1988), 64 O.R. (2d) 289, 3 C.E.L.R. (N.S.) 330 (H.C.).	4.3.3
Chan v. Rowley, [1992] B.C.W.L.D. 1265 (S.C.).	2.2.1; 2.3.2 4.3.3

Case	Section
Choppin v. Caouette (1982), 41 A.R. 32 (Q.B.).	1.2.3
C.K.O.Y. Ltd. Ottawa (City), [1952] O.W.N. 367 (Co. Ct.).	5.3
Cloutier v. Ball, [1995] B.C.J. No. 1301 (S.C.).	8.2
Cordiez v. Muffly, unreported, November 30, 1991, Rossland No. 6041 (B.C.S.C.).	2.2.2; 3.3 3.5
Corporation Municipale D'Ulverton v. Brock et al (1983), 23 M.P.L.R. 109 (Que. S.C.).	9.3
Cosco v. Scott, unreported, November 28, 1984, Nanaimo No. 6041 (B.C. Co. Ct.).	3.6
Crawford v. Crawford, [1986] N.B.J. No. 771 (Q.B.).	2.2.1
Crosty v. Burlington (City) (1978), 21 O.R. (2d) 753, (1978) 91 D.L.R. (3d) 572 (H.C.).	1.2.2; 4.3.1
Crowhurst v. Amersham Burial Board, [1874] 4 Ex. D. 5.	4.2
Cunliffe v. Bankes, [1945] 1 All E.R. 459 K.B.).	5.3
Davey v. Harrow Corporation, [1958] 1 Q.B. 60 (C.A.).	4.3.1
Davidson v. Kamloops (City), [1992] B.C.J. No. 1010, (1992)10 M.P.L.R. (2d) 276, [1992] C.C.L. 9738 (S.C.); affirmed [1995] B.C.J. No. 435,[1995] 56 B.C.A.C. 201, 92 W.A.C. 201 (C.A.).	5.2
Dixon v. Patton, [1986] A.J. No. 821, (1986) 47 Alta. L.R. (2d) 76, 75 A.R. 83 (Q.B.).	7.2
Doucette v. Parent, [1996] O.J. No. 3493 (C.A.).	8.4; 8.5
Dykeman (Guardian ad litem of) v. Emerson, [1991] B.C.J. No. 198 (S.C.).	8.3
Dykhuizen v. Saanich (District), [1989] B.C.W.L.D. 2660, (1989) 63 D.L.R. (4th) 211 (B.C.C.A.).	2.3.2; 3.4.3
Ficht v. St. Catherines (City) (1978), 23 O.R. (2d) 229, 94 D.L.R. (3d) 703 (H.C.).	5.2
Forbes v. Griffin (1989), 99 N.B.R. (2d)	3.4.2

Case	Section
Homes by Jayman Ltd. v. Kellam Berg Engineering & Surveys Ltd. (1995), 29 Alrta. L.R. (3d) 1, (1995) 20 C.L.R. (2d) 161, (1995) 171 A.R. 1, [1995] A.W.L.D. 577, [1995] A.J. No 471 (Q.B.).	10.5.1
Horseshoe Bay Retirement Society v. S.I.F. Development Corp. (1990), 66 D.L.R. (4th) 42, (1990) 3 C.C.L.T. (2d) 75, [1991] B.C.W.L.D. 1745 (S.C.).	2.2.1; 2.2.2 2.3.2; 3.4.3
Horton v. Speta, [1996] B.C.W.L.D. 1529; additional reasons [1996] B.C.W.L.D. 1846 (S.C.).	3.6
Howard v. Morgan Trust (1983), 25 C.C.L.T. 64 (Que. S.C.).	2.2.1; 3.4.3 3.6
Huestis v. Toronto (City) (1926), 58 O.L.R. 648, [1926] 3 D.L.R. 142 (C.A.).	1.2.2; 5.3
Hugh v. Vancouver (City), [1981] 5.W.W.R. 250, (1981) 126 D.L.R. (3rd) 527 (B.C.S.C.).	5.5
Jordan House v. Menow and Honsberger, [1974] 1 S.C.R. 239, (1974) 38 D.L.R. (3d) 105.	4.3.1
Just v. British Columbia, [1989] 2 S.C.R. 1228, 41 B.C.L.R.(2d) 350, (1989) 64 D.L.R. (4th) 689, [1990] 1 W.W.R. 385, 1 C.C.L.T. (2d) 1, 18 M.V.R. (2d) 1; affirming 10 B.C.L.R. (2d) 223, [1987] 2 W.W.R. 231, 40 C.C.L.T. 160 (C.A.); affirming 64 B.C.L.R. 349, [1985] 5 W.W.R. 570, 34 M.V.R. 124 (S.C.).	5.3
Kahl v. Oak Bay (District), unreported, May 1, 1978, Victoria Registry No. 1711/77 (B.C.S.C.).	8.6
Kamin v. Windsor (City), unreported, June 19, 1990, Windsor File No. D-3241/89 (Ont. Prov. Ct.).	5.3
Kates v. Hall (1990), 46 C.C.L.T. 186, (1990) 47 B.C.L.R. (2d) 329, [1990] 5 W.W.R. 569, [1990] B.C.W.L.D. 1874 (S.C.).	1.3

Case	Section
Kates v. Hall (1991), 53 B.C.L.R. (2d) 322, [1991] B.C.W.L.D. 617 (C.A.); affirming [1989] B.C.J. No. 1358, [1989] B.C.W.L.D. 2048 (S.C.).	2.2; 2.3.2 3.4.3
Kelley v. Ingersoll (Town), [1996] O.J. No. 2318, [1996] 33 M.P.L.R. (2d) 100 (Ont. Gen. Div.).	1.2.1; 8.4
Leakey v. Nat. Trust for Places of Historic Interest or Natural Beauty, [1980] 1 Q.B. 485, [1980] 2 W.L.R. 65, [1980] 1 All E.R. 17 (C.A.); affirming [1978] Q.B. 849, [1978] 3 All E.R. 234.	1.2.1; 8.4
Lee v. Hersh, [1993] B.C.W.L.D. 1138 (S.C.).	2.2.1; 3.4.3
Leischner v. West Kootenay (1986), 70 B.C.L.R. 145, [1986] 3 W.W.R. 97, (1986) 24 D.L.R. (4th) 641 (C.A.); reversing in part (1981), 130 D.L.R. (3d) 373 (B.C.S.C.); varied on reconsideration (1983), 45 B.C.L.R. 204, [1983] 5 W.W.R. 166, 35 C.P.C. 151, 150 D.L.R. (3d) 242 (S.C.).	6.3
Lemmon v. Webb, [1894] 3 Ch. 1 (C.A.); affd. [1895] A.C. 1 (H.L.).	4.2
Loverock v. Webb (1921), 30 B.C.R. 327, 70 D.L.R. 748 (C.A.).	4.3.3
MacDonald v. Goderich (Town), [1947] O.R. 908, [1948] 1 D.L.R. 11 (C.A.).	6.3
MacKenzie v. Vancouver Lawn Tennis and Badminton Club, [1978] B.C.J. No. 325 (S.C.).	8.4
Marshak v. Toronto (City), [1953] O.W.N. 101 (H.C.).	5.5
McCombe v. Read, [1955] 2 Q.B. 429, [1955] 2 All E.R. 458.	4.3.1
McKim v. Degraaf, [1991] B.C.J. No. 2841 (S.C.).	3.4.3
Melito v. Lenbro Holdings Ltd. (1983), 13 C.E.L.R. 37 (Ont. Co. Ct.).	7.2
Mendez v. Palazzi (1976), 12 O.R. (2d) 270, 68 D.L.R. (3d) 582 (Co. Ct.).	4.3.1
Merest v. Harvey [1814] 5 Taunt. 442, 128 E.R. 761.	2.2.2

Case	Section
Michaels v. Zwickel, [1994] B.C.W.L.D. 918 (S.C.).	2.2.1; 4.3.3
Millar & Brown v. Vancouver (City) (1966), 58 W.W.R. 191, 59 D.L.R. (2d) 640 (B.C.C.A.); reversing (1965), 55 W.W.R. 59, 56 D.L.R. (2d) 190 (B.C.S.C.).	5.3
Miller v. Kamloops (City), [1996] B.C.J. No. 1496 (S.C.).	8.4
Milliken v. Kidd and Kidd, unrep., March 18, (1994), File 1629/93 (Ont. Gen. Div. - Sm.Cl.).	1.2.3; 2.3.2
Mills v. Brooker, [1919] 1 K.B. 555.	4.2
Mills v. MacAleese (1987), 83 N.B.R. (2d) 292, 212 A.P.R. 292 (C.A.).	3.4.1; 3.6
Morris v. Fraser (1965), 53 W.W.R. 1275, (1924) 33 B.C.R. 474, [1924] 2 D.L.R., 811	6.3
Moule v. New Brunswick Electric Power Commission (1960), 44 M.P.R. 317, 81 C.R.T.C. 265, 24 D.L.R. (2d) 305 (S.C.C.); affirming (1959), 44 M.P.L.R. 310, 22 D.L.R. (2d) 253 (N.B.C.A.).	6.3
Muncaster v. Nunnemacher, [1994] B.C.W.L.D. 1701 (S.C.).	2.2.1
Muncaster v. Nunnemacher, [1996] B.C.J. No. 885, (1996), 76 B.C.A.C. 211, (1996) 125 W.A.C. 211, [1996] B.C.W.L.D. 1340 (C.A.).	2.2.1
Nailen v. Powell River (Municipality), [1978] W.W.R. 350 (B.C. - Co. Ct.).	3.3; 3.5
National Moving & Storage Ltd. v. Fredericton (City) (1983), 27 C.C.L.T. 286, 51 N.B.R. (2d) 303, 134 A.P.R. 303 (Q.B.).	5.3
Neal v. West Kootenay Power Ltd., [1993] B.C.J. No. 1667 (S.C.).	6.3
Noble v. Harrison, [1926] K.B. 284, [1926] All E.R. 284.	5.3

Case	Section
O'Brien v. Underwood McLellan & Associates Ltd. (1979), 5 Sask. R. 337 (Q.B.).	2.2.1
O'Cadlaigh v. Madiuk, [1994] B.C.W.L.D. 1573, additional reasons (1994) C.P.C. (3d) 116, [1994] B.C.W.L.D. 2944 (S.C.).	2.2.1; 2.2.2 3.4.3
Oran v. Westwood Fibre Ltd., [1996] B.C.J. No. 2697 (S.C.).	2.3.2
Ottawa Electric Co. v. Crepin, [1931] S.C.R. 407, 38 C.R.C. 44, [1931] 3 D.L.R. 113; affirming (1930), 66 O.L.R. 409, [1931] 1 D.L.R. 792 (C.A.).	6.3
Patterson v. N. Vancouver Board of School Trustees (1928), [1929] 2 W.W.R. 181, 41 B.C.R. 123, [1929] 3 D.L.R. 33 (C.A.).	8.4
Peters v. Diamond, [1964] 1 O.R. 139, (1964) 41 D.L.R. (2d) 311 (Co. Ct.).	2.3.2; 3.3
Peters v. Dodge (1910), 45 N.S.R. 33 (C.A.).	4.3.3
Petrillo v. Ikotin, [1993] M.J. No. 252 (C.A.).	3.3
Pretu v. Donald Tidey Co. Ltd., [1966] 1 O.R. 191, (1966) 53 D.L.R. (2d) 504 (H.C.); (1966) 53 D.L.R. (2d) 509n. (Ont. C.A.).	2.2.2
Prince Rupert (City) v. Pederson (1994), 98 B.C.L.R. (2d) 84, [1995] 1 W.W.R. 421, [1994] B.C.W.L.D. 2663, (1994) 2 D.M.P.L. 591 (C.A.); varying [1992] B.C.W.L.D. 848, (1992) 1 D.M.P.L. 495 (S.C.).	2.2.1; 2.2.2 3.4.3
Purdy v. Kneip, [1975] 3 W.W.R. 573 (B.C.S.C.).	8.2
R. v. Abbey, [1982] 2 S.C.R. 24, 68 C.C.C. (2d) 394, 29 C.P.R. (3d) 202, [1983] 1 W.W.R. 251, 39 B.C.L.R. 201, 43 N.R. 30.	10.9
R. v. Bata Industries Ltd. (1992), 9 O.R. (3d) 329p, (1992) 70 C.C.C. (3d) 394 p, (1992) 7 C.E.L.R. (NS) 245, (1992) 1 D.E.L.E.A. 107 (Prov. Div.); varied [1993] O.J. No. 1679, (1993) 14 O.R. (3d) 354, (1993) 11 C.E.L.R. (NS) 208 (Gen. Div.);	10.13.2

Case	Section
Scott v. City of Nanaimo, [1924] 1 W.W.R. 1275, (1924) 33 B.C.R. 474, [1924] 2 D.L.R. 811 (C.A.).	5.5
Sharp v. Powell (1872), L.R. 7 C.P. 253, 41 L.J.C.P. 95, 26 L.T. 436, 20 W.R. 584.	6.3
Sinkewicz v. Schmidt, [1994] S.J. No. 91, [1994] 4 W.W.R. 569, (1994) 118 Sask. R. 112 (Q.B.).	1.4; 2.3.1
Smith v. Giddy, [1904] 2 K.B. 448, 73 L.J.K.B. 894.	4.2
Steves v. South Vancouver (District) (1897), 6 B.C.R. 17 (F.C.).	5.3
Stockinger v. Coburg (Town), [1944] O.R. 372, [1944] 3 D.L.R. 104 (C.A.); affirming [1943] O.W.N. 461, [1943] 4 D.L.R. 357 (H.C.).	5.4
Stowe v. McTaggart, [1993] B.C.J. No. 1815 (Prov. Ct.).	2.2.1; 4.3.2 4.3.3
Swinamer v. Nova Scotia (Attorney-General) (1992), 19 C.C.L.T. (2d) 233, 2 M.V.R. (3d) 80, 163 N.R. 291, [1994] 1 S.C.R. 445, 129 N.S.R. (2d) 321, 362 A.P.R. 321, 112 D.L.R.(4th) 18, 20 Admin. L.R. (2d) 39; affg. (1992) 35 M.V.R. (2d) 191, 10 C.C.L.T. (2d) 207, 108 N.S.R. (2d) 254, 294 A.P.R. 254 (S.C. - A.D.); reversing (1991), 6 C.C.L.T. (2d) 270, 101 N.S.R. (2d) 333, 275 A.P.R. 333 (T.D.).	5.3
Symons v. Madden, [1987] B.C.J. No. 2620 (Co. Ct.).	3.6
Timpauer v. Coffey, unreported, September 21, 1991, Vancouver No. C900223 (B.C.S.C.).	1.3; 2.2.1
Thornhill v. Martineau, [1987] B.C.J. No. 467, (1987) 39 C.C.L.T. 293, 2 M.V.R. (2d) 141 (S.C.).	5.2
Urquhart v. Sewell, [1991] B.C.J. No. 3449 (S.C.).	3.4.3
Ursel v. Jenkins, [1988] B.C.J. No. 1277 (Co. Ct.); affirmed *(sub nom. Ursel v. Van Schilt)*, [1989] B.C.J. No. 775 (C.A.).	3.4.3; 3.5

Case	Section
Valentine Lands Corp. v. Gibsons (Town), [1992] B.C.J. No. 272, (1992) 9 M.P.L.R. (2d) 69 (S.C.).	9.3
Vancouver Community College v. Phillips Barrett (1987), 20 B.C.L.R. (2d) 289, (1987) 27 C.L.R. 11, (1987) 38 L.C.R. 30, [1988] B.C.W.L.D. 371 (S.C.).	10.7.3
Verkerk v. McDowell, [1996] B.C.W.L.D. 2315 (S.C.).	3.6
Wahl v. Listowel (Town) (1988), 40 M.P.L.R. 276 (Ont. D.C.).	5.3
Webb v. Attewell, [1994] 4 W.W.R. 404 (B.C.S.C.).	3.2
Whistler Housing Corp. v. Whistler (Resort Municipality) (1994), B.C.L.R. (2d) 62, [1994] B.C.W.L.D. 912, (1994) 2 D.M.P.L. 427 (S.C.).	9.3
Willard G. Hallman Lumber Ltd. v. Grey (County) (1985), 15 Admin L.R. 224 (Ont. S.C.).	9.3
Winger v. Torkelson, [1989] O.J. No. 2643 (Prov. Ct.).	4.3.2; 4.3.3
Wolverine Tube v. Noranda, [1995] O.J. No. 3529, (1995) 26 O.R. (3d) 577, (1995) 87 O.A.C. 191, (1995) 18 C.E.L.R. (N.S.) 219, (1995) 6 W.D.C.P. (2d) 548, (1995) 5 D.E.L.E.A. 7 (C.A.); affirming [1994] O.J. No. 2696, (1994) 21 O.R. (3d) 264, (1994) 4 D.E.L.E.A. 10 (Ont. Gen. Div.).	10.7.4
Yadlos v. Sturdy, [1989] B.C.J. No. 2191 (Co. Ct.).	7.4
Young v. Toronto (City) (1985), 31 M.P.L.R. 142 (Ont. D.C.)	9.4
698419 Ontario Ltd. v. Biggs, unreported, February 10, 1994, Simcoe 353/93 (Ont. Gen. Div.).	1.4; 3.4.3

Notes

Note	Chapter #
[1] This case appears to be a classic example of the difference between policy and implementation. The Town's policy was to implement the tree inspection, but the actual manner of implementation was found wanting, thus demonstrating negligence and inviting liability.	8
[2] A useful review of Ontario cases can be found in Westley, 1995.	9
[3] The authors gratefully acknowledge the assistance and input of J.M. (Tim) Mackenzie in preparing this chapter.	10
[4] Sopinka, Lederman and Bryant, (1992)	10
[5] Sopinka, Lederman and Bryant, *supra*, n. 4	10
[6] From an unpublished paper prepared by G.A. Urquhart, Q.C. entitled "Complex Litigation - Experts."	10
[7] The exact specifications for provincial court rules may vary across Canada. Consult a lawyer before finalising this aspect of your report.	10

References

Abeyta, D. (Ed.) 1995. A Guide to Report Writing for Consulting Arborists. Savoy: International Society of Arboriculture.

American Society of Consulting Arborists. 1996. Code of Ethics & Standards of Professional Practice. Rockville, Maryland: ASCA.

Bernhardt, E.A. and T.J. Swiecki. 1991. Guidelines for Developing and Evaluating Tree Ordinances. Sacremento: California Department of Forestry and Fire Protection.

Black, H.C. 1990. Black's Law Dictionary 6th Ed. St. Paul, Minnesota: West Publishing Co.

Dunster, J.A., and K. Dunster. 1996. Dictionary of Natural Resource Management. Vancouver: University of British Columbia Press.

Feder, H.A. 1993. Succeeding as an Expert Witness. Glenwood Springs, Colorado: Tageh Press

Fridman, G.H.L. 1990. The Law of Torts in Canada. Volume 2. Toronto: Carswell.

Hendricksen, J., "Conscious Incompetency," in (1996)2(4):3 *Certification Today.*

Hoefer, P.J. Himelick, E.B. and D.F. De Voto. 1990. Municipal Tree Manual. Savoy: International Society of Arboriculture.

Municipal Act, R.S.B.C. 1996, c. 323.

Sopinka, J., S.N. Lederman, Q.C. and A.W. Bryant. 1992. The Law of Evidence in Canada. Toronto: Butterworths

Trees Act, R.S.O. 1990, c. T-20

Westley, K. 1995. Municipal Tree Cutting By-Laws (Passed Pursuant to the Trees Act) Case Inventory. Midhurst: County Clerk, Simcoe, Ontario.

Wright, C.A., A.M. Linden, and L.N. Klar. 1985. Canadian Tort Law. Cases, Notes and Materials. 8th Ed. Toronto: Butterworths.

About the Authors

Dr. Julian Dunster is a Registered Professional Forester, a Registered Planner (Canada and British Columbia), an ISA Certified Arborist, and a candidate member in the American Society of Consulting Arborists. He holds a doctorate in Regional Planning and Resource Development from the University of Waterloo, and Masters and Bachelors degrees in Forestry from the University of British Columbia and Bangor University in North Wales, Great Britain. Julian has taught at several universities and, during the course of research and consulting assignments, has delivered guest lectures in many parts of the world. He has published over thirty articles on forestry and environmental planning issues, and is the co-author of the Dictionary of Natural Resource Management (UBC Press 1996). He is President of an environmental consulting company based on Bowen Island, British Columbia, Canada.

Ms. Susan Murray is a Registered Professional Agrologist with memberships in the British Columbia Institute of Agrologists and the Canadian Institute of Agrologists, and is a Certified Arborist with the International Society of Arboriculture. She received her Master of Science and Bachelor of Science degrees in Agriculture from the University of British Columbia. She is an experienced educator and has taught horticulture and arboriculture at the post-secondary level since 1979. She currently teaches at Kwantlen University College in Langley, British Columbia. She also owns and operates an arboricultural consulting firm.